ARTILLERY

ARTILLERY

FROM THE CIVIL WAR TO THE PRESENT DAY

Michael E. Haskew

METRO BOOKS
NEW YORK

Metro Books
122 Fifth Avenue
New York, NY 10011

Editorial and design by
Amber Books Ltd

Project Editor: Sarah Uttridge
Picture Research: Terry Forshaw and Kate Green
Design: Graham Curd

ISBN 13: 978-1-4351-0124-1
ISBN 10: 1-4351-0124-3

Printed and bound in China

10 9 8 7 6 5 4 3 2 1

Picture acknowledgements:
All photographs courtesy of Art-Tech/Aerospace except for the following:
Art-Tech/MARS: 29(r), 33(r), 71, 85, 96, 98, 165, 169, 181
BAE Systems: 206, 214, 219
Cody Images: 11, 12, 26, 38, 57, 61, 112-114, 152, 201
Corbis: 9, 151(both), 200, 218(John O'Boyle/Star Ledger)
Mary Evans Picture Library: 19-21, 24, 25
Getty Images: 16, 144
Library of Congress: 6, 13-15, 17, 18, 36(r)
McGill University: 210
Photos.com: 10, 22, 46, 49
Popperfoto: 33(l)
Rex Features: 211 (Sipa Press); 212 (Simon Walker)
Suddeutscher Verlag: 47
ThyssenKrupp AG: 52
Ukrainian State Archive: 75, 98, 115, 120
US Air Force: 178
US Department of Defense: 32, 34, 41, 43, 58, 64, 70(r), 78, 140, 166, 175,
180, 183-186, 188, 190, 191, 194, 195, 197, 205, 215

All artworks courtesy of Art-Tech/Aerospace except for the following:
Art-Tech/John Batchelor: 8, 29(l)

CONTENTS

Introduction   6

World War I   26

The Interwar Years   62

Victory in Europe and the Pacific   98

The Cold War   140

Modern Artillery   186

Glossary   220
Index   221

INTRODUCTION

The impact of the Industrial Revolution on the technology of war became obvious in the mid-nineteenth century as artillery rapidly advanced in size, durability and destructive power. Innovations such as the development of mass production, improved metallurgy and the rise of the factory fostered an unprecedented period of change.

Well documented in military history is the capacity of artillery to inflict casualties and demoralize an enemy – hence its nickname, the 'queen of the battlefield'. Artillery has, on numerous occasions, paved the way for victory or saved a crumbling army from what otherwise might have been a devastating defeat. During the last 150 years, artillery has developed to the extent that it has decided the outcome of a battle and, inevitably, the course of history.

Artillery is probably as old as humanity. After all, the hostile throwing of a stone is, in itself, the employment of additional firepower, utilizing an object of mass and the force of gravity to inflict damage upon an opponent. Centuries before the birth of Christ, Philip of Macedon was said to have considered primitive artillery an essential component of his army. Philip's army employed the

On the morning of 2 January 1863, a desperate Confederate charge was shattered at the Battle of Stones River by the combined fire of more than 50 Union cannon.

gastrophete, the equivalent of an oversize crossbow, which fired a large arrow-like projectile at the release of a tension line. Another weapon in King Philip's arsenal was the palintonon, which was capable of hurling a stone weighing 3.6kg (8lb) or more a distance of up to 274m (300yds).

While Philip probably used these primarily as weapons of siege warfare, it was his son, Alexander the Great, whose artillery took on a more decidedly offensive role. Alexander's artillery was often used to provide covering fire for advancing infantry, and further developments allowed lighter, more mobile forms of artillery to be carried by the troops and assembled on the battlefield.

The ancient historian Josephus documented the employment of the onager by Roman legions during the siege of Jerusalem. Literally translated as 'wild ass' due to the substantial recoil of the weapon, the onager could throw a 45kg (100lb) stone a distance of up to 365m (400yds). These siege weapons eventually became standard components of the legions of the Roman Empire.

Other forms of Roman artillery included the ballista, which was easily transported and could fire a 4.5kg (10lb) stone 274m (300yds), and the carroballista, which hurled a projectile 66cm (26in) in diameter a distance of 274m (300yds).

Obviously, the use of artillery during a siege operation was intended to pulverize the fortifications of a city and to demoralize the defending army and civilian population. In open battle, the mere presence of imposing weaponry could decidedly impact the enemy's will to fight. If used effectively, artillery was capable of inflicting relatively heavier casualties than close combat at something of a standoff distance.

The first employment of artillery that used gunpowder as a propellant is said to have occurred in China during the period of the Soong Dynasty of the late twelfth century. Large, smoothbore cannon called bombards were used by the Turks during the siege of Constantinople in the mid-fifteenth century. These massive weapons were constructed of metal rods bound together,

weighing 21 tonnes (19 tons), and requiring 60 oxen and a crew of 200 men to position them. The word 'cannon' became familiar during the Hussite Wars of Bohemia (1418–24), as Jan Zizka deployed artillery on wheeled carriages transported by hauling animals.

Inherent difficulties existed with early artillery, particularly the transportation and deployment of individual pieces for battle. Rates of fire could be ponderously slow and fail to decisively affect the outcome of a battle. Over the next 200 years, innovations such as the combination of propellant and projectile into a single unit, or cartridge, and the development of fused explosive shells contributed to a resurgence in the use of artillery by the world's armies and navies. The Swedish King Gustavus Adolphus was the principal force behind the development of smaller and lighter artillery, such as the three-pounder gun used by his troops during the Thirty Years' War of the early 1600s.

The premiere treatise on the use of artillery was written during this period by Kazimierz Siemienowicz, entitled *Great Art of Artillery, the First Part*. The essential elements of artillery, its deployment on the battlefield and in static defensive positions, and its basic composition were to change little for more than 200 years.

Emergence of the modern era

With the dawn of the Industrial Revolution, remarkable advances in the development of artillery occurred within a relatively short space of time. The concentration of workers in factories, dedicating a labour force to specific production, and the spreading concept of interchangeable parts, developed by the American Eli Whitney, resulted in a great leap forward in productive capacity. The introduction of steam to power machinery and transportation, as well as the emergence of the railway as a prime mover of military forces and the

Large smoothbore cannon like the bombard were used as siege weapons. One notable occurrence was at the Siege of Constantinople in the fifteenth century.

logistical apparatus necessary to support larger armies, influenced wars of imperialism and empire.

Advances in the development of artillery occurred in tandem with other military innovations during the mid-nineteenth century. Rifled small arms, and later repeating weapons and machine guns, dramatically increased the firepower and lethality of infantry. Warships of iron rendered the world's navies of wooden sailing ships obsolete virtually overnight. Experimentation with aircraft and submarines heralded an era of modern warfare, which came to deadly fruition in World War I, between 1914 and 1918.

Innovations in artillery design and deployment had remained generally stagnant for 300 years, and then from about 1850 the accelerated pace forward was astounding. The most significant innovations included the conversion of the weapons themselves from smoothbore, muzzle-loading pieces to rifled barrels serviced as breech-loaders. The introduction of streamlined projectiles with various types of timed fuses; more efficient propellants such as cordite, introduced by the British on the eve of World War I; the development of recoil systems, which improved rates of fire and relieved crews of the arduous task of wrestling a weapon back into firing position following each round; the common use of ammunition encompassing the projectile, propellant and fuse in one piece; and the employment of the shrapnel shell, which sprayed steel fragments in all directions upon detonation – all fostered the realization among military strategists and tacticians that concentrated, well-commanded artillery could be devastating and decisive.

The manufacture of rifled artillery capable of handling larger projectiles and propellant charges presented challenges in durability. In 1854, during the Crimean War, Sir William Armstrong, a British hydraulic engineer, introduced a method of forging barrels from wrought iron, first twisting iron bars and then welding them together through a forging process. An improvement on the original design was accomplished with a steel gun tube reinforced by wrought-iron coils. Armstrong had previously founded a manufacturing company and produced artillery in several sizes. One of the best known was the 12-pounder Armstrong rifle, with a bore of 7.6cm (3in) and a screw mechanism, which closed

Heavy equipment is used to hoist the cast iron behemoth of a heavy cannon from the floor of an American factory during the mid-nineteenth century.

Sir William Armstrong, a British engineer, was a pioneer in the development of artillery through improved manufacturing techniques.

the breech. The Armstrong 12-pounder was capable of firing its projectile a distance of 8340m (9175yds), or 8.4km (5.2 miles). Eventually, the Armstrong weaponry was adopted as standard by both the British Army and the Royal Navy.

Armaments manufacturers in other countries were producing improved artillery as well. The Frenchman Treuille de Beaulieu designed a muzzle-loading rifled cannon, while the Krupp family of Germany manufactured a breech-loader from a single billet of steel. Innovators struggled with maintaining pressure for propulsion inside the breech, and various methods emerged of sealing the breeches with screw mechanisms, expanding metal rings or the sliding breechblock developed later by Krupp.

By this time, the advantages of rifling – the cutting of spiral grooves into the barrel – were being actively applied to infantry weapons. Moreover, the principles of longer-range, greater-accuracy and more highly concentrated penetrating power were proven uniform in heavier weapons. With increased range, the concept of indirect fire became a standard process in the latter part of the nineteenth century. Artillery crews routinely fired on distant targets that could not necessarily be observed with the naked eye.

Breech-loading allowed for an improved rate of fire and also meant that the crews serving the artillery pieces did so with less exposure to the enemy. During the Crimean War (1853–56), the British Army fielded its first breech-loading cannon. Subsequently, the Russo-Turkish War of 1877–78 was the first war in which artillery and infantry were simultaneously equipped with rifled breech-loading cannon and repeating small arms.

One of the earliest examples of the devastating effect of massed artillery occurred during the

THE CRIMEAN WAR

Acknowledged by many historians as the first modern war, the Crimean War involved the introduction of modern, breech-loading artillery with the British Army and also the first use of the telegraph as a means of communication in wartime, extensive employment of railroads for transportation, trenchlines for the protection of troops, and the introduction of the Minié ball as rifle ammunition. Primitive photography also produced the first such images of the battlefield, while modern nursing techniques were developed by pioneers such as Florence Nightingale.

Crimean War on 25 October 1854, at the Battle of Balaclava. The epic charge of the Light Brigade, immortalized in the poem of the same name by Alfred, Lord Tennyson, recounted:

Half a league, half a league,
    Half a league onward,
All in the valley of Death
    Rode the six hundred.
'Forward, the Light Brigade!
Charge for the guns!' he said:
    Into the valley of Death
    Rode the six hundred…

    Cannon to right of them,
Cannon to left of them,
    Cannon in front of them
Volley'd and thunder'd
Storm'd at with shot and shell,
    Boldly they rode and well
Into the jaws of Death,
    Into the mouth of Hell
Rode the six hundred…

When the charge was over, the combined Russian artillery and rifle fire had ravaged the British cavalry. Of a complement numbering 673 men, 113 were dead, 134 wounded and as many as 500 horses had been shot down.

The rapid development of artillery gave rise to a well-defined classification of the weapons used. Three broad terms – guns, howitzers and mortars – have generally described these implements of warfare for the last century-and-a-half. Guns have usually been defined as weapons with long barrels for longer range, maximum elevations of less than 45° resulting in a flatter trajectory of fire, and ammunition encasing propellant, projectile and fuse in one. Howitzers are classified as weapons that are capable of greater plunging fire, with trajectories ranging from 45° and greater. Early howitzers are distinguished from guns by their shorter barrels and lower muzzle velocities. Through the years, however, the distinction between guns and howitzers has lessened considerably, particularly with the development of weapons capable of a wide range of elevations and of firing numerous types of ammunition. Mortars, depending on size, crew and ammunition requirements, may sometimes be properly classified as small arms. However, large versions of mortars have been employed as siege weapons or during prolonged engagements. Most mortars are smoothbore muzzle-loading weapons, which fire at a high trajectory.

The crucible of the Civil War

The American Civil War (1861–65) is often referred to as the first truly modern war, given that it served as a proving ground for much of the enhanced technology that had recently been developed. During the Civil War, the artillery of both the Union and the Confederacy depended upon readily available smoothbore artillery, such as the Model 1857 12-pounder Napoleon, the last of a number of guns cast in bronze to be used by an army of the United States. The Napoleon was a reliable weapon and was deadly when firing canister or grapeshot against massed bodies of enemy troops.

Named after Emperor Napoleon III of France, the Model 1857 12-pounder could fire a 5.6kg (12.3lb) projectile a distance of just over 1463m (1600yds). Its bore was 11.7cm (4.62in), and its barrel length 1.67m (5.48ft). The Napoleon was sometimes referred to as a 'gun-howitzer' because it was considered by many to have characteristics of both.

In comparison, Union forces possessed a great many more rifled cannon than the Confederates, who suffered from continual logistical shortages due to blockades, Union occupation of Confederate territory, and manpower shortages.

HEAVY BRITISH CANNON

On display as a museum piece and still seated on a portion of its wooden gun carriage, this cannon of British manufacture is similar to those deployed with the British Army during the Crimean War (1853–56), including the initial breech-loading weapons placed in service. Concentrated firepower from Russian artillery decimated the ranks of the Light Brigade in the Crimea.

As the war progressed and the supply of replacement artillery from sources outside the Confederacy dwindled to virtually nothing, foundries and arms-manufacturing facilities that had operated in the Deep South came under Union occupation as well.

The 3in ordnance rifle was produced in greater numbers than any other rifled artillery piece of the Civil War and was renowned for its accuracy. 'The Yankee three-inch rifle was a dead shot at any distance under a mile,' commented one Confederate gunner who endured counter-battery fire from the weapon during the Atlanta Campaign of 1864. 'They could hit the end of a flour barrel more often than miss, unless the gunner got rattled.' Indeed, it was often the mission of the sharpshooters of both sides to target artillerymen who serviced the guns of their enemies. Cannon without crews were a considerably lower threat than those that were fully operational.

Designed by inventor John Griffen, the iron 3in rifle fired a 4.3kg (9.5lb) projectile a distance of 1674m (1830yds). Its barrel length was 175cm (69in) and it weighed a relatively light 372kg (820lb). Other common artillery pieces of the Civil War included the 12- and 24-pounder howitzers, and the 10- and 20-pounder Parrott rifles. Invented by Robert Parker Parrott, the Parrott rifles were manufactured in a number of sizes, including a colossal 300-pounder. However, most of the guns to bear Parrott's name and use the ammunition he designed were produced in the smaller sizes. Both Union and Confederate forces employed these weapons regularly.

The 12-pounder Napoleon, named for Emperor Napoleon III of France, was commonly employed during the American Civil War.

Parrott's cannon, like others, were known to fail under the rigours of prolonged combat. Sometimes the weapons exploded during use, killing or injuring the artillerymen servicing them. In a letter dated 15 November 1864, Parrott responded to concerns raised by an officer of engineers during the siege of Charleston, South Carolina, by Union forces the previous year:

'I believe that when we consider the precautions which experience has proved to be proper in the use of rifle cannon; the effect of the premature explosion of shells, and the exposure of these guns to drifting sands, there will be little left to weigh against the evidence so generally given in favor of the safety of the guns themselves… In so new a field of enterprise, accidents, not necessarily connected with rifled ordnance, but the result of want of experience, were to be expected. That my guns have encountered difficulties (now as I think remedied) because so early in that field, so largely called for and used, cannot justly operate to their prejudice.'

Of particular interest among Civil War artillery is the 12-pounder Whitworth breech-loading rifle.

In this romanticized view of the Battle of Kennesaw Mountain, Georgia, Union soldiers are depicted charging directly into the firing muzzles of Confederate cannon.

This was designed by Sir Joseph Whitworth, a British mechanical engineer, and no fewer than five muzzle-loading and seven breech-loading variants of the Whitworth rifle were used during the Civil War. In 1863, during testing at Southport Sands, England, it was reported that a breech-loading Whitworth fired a projectile that actually hit its target nearly 8km (5 miles) away. The breech-loading Whitworth was relatively rare during the Civil War, but in 1864, an article in a magazine on the topic of engineering validated its reputation for incredible accuracy: 'At 1600 yards the Whitworth gun fired 10 shots with a lateral deviation of only 5 inches.'

Perhaps the most significant aspect of the breech-loading Whitworth during the Civil War was that it constituted a precursor to the standardization of the world's great armies during the following decades. The weapon was said to have made a distinctive report when fired due to the fitting of the boltlike projectile into the grooves of the hexagonal bore.

A harvest of death

The terrible devastation wrought by artillery during the Civil War is starkly evidenced by photographs of battlefields taken soon after the fighting had ended and prior to the arrival of burial details. Mangled, disembowelled corpses lie strewn across the ground in gruesome poses. Various types of ammunition were used, depending upon the demands of the tactical situation. Among these were solid shot, spherical shells, which would explode over the heads of the enemy and rain down fragments, and case projectiles, such as the canister, which were loaded with iron balls to shower troop formations in a similar fashion to massed rifle fire.

Canister was sometimes fired in double and was deadly up to a range of 366m (400yds), essentially turning the cannon into a large shotgun. A thin container holding nearly 30 iron balls cushioned by sawdust disintegrated when the weapon was discharged and allowed the balls to blast forward with devastating effect. An earlier form of canister, called grapeshot, also held such anti-personnel balls. In addition to canister and grapeshot, other types of deadly fragmentation shells exploded into jagged pieces of iron, which could be deadly to massed troops in the open. Named for General Henry Shrapnel, the British officer and inventor who pioneered their development and use, these shells raised battlefield casualty levels significantly.

During the Civil War, Union and Confederate batteries often engaged in counter-battery fire in an attempt to establish command of the field and facilitate the movement of infantry. One observer wrote of a Confederate battery that had been ranged and the devastated by Union artillery:

'Such a scene as it presented – guns dismounted and disabled, carriages splintered and crushed, ammunition chests exploded, limbers upset, wounded horses plunging and kicking, dashing out the brains of men tangled in the harness; while cannoneers with pistols were crawling around through the wreck shooting the struggling horses to save the lives of wounded men.'

The use of horse-drawn limbers, caissons and field pieces – hence the term 'horse artillery' or 'flying artillery' – was critical to the rapid movement of firepower to areas of immediate need during the Civil War. Some 20 years earlier, the concept of flying artillery had been employed successfully by the US Army during the war with Mexico (1846–48). The result was that now many of the senior commanders on both sides of the present conflict understood the value of such mobility. Aerial observation of enemy troop dispositions and movements took place for the first time in military history as Thaddeus Lowe ingeniously devised a system to inflate large balloons, which were then allowed to rise on tethers to an optimal height of 91m (300ft). Lowe himself relayed information on Confederate dispositions from aloft during early campaigns.

General Gibbon at Antietam

Before the war, Union General John Gibbon had earned a commission as an artillery officer after graduating from the United States Military

Academy at West Point, New York, in 1847. At the height of the Battle of Antietam in September 1862, Gibbon observed Battery B of his old unit, the 4th US Artillery, in action, trying to save its threatened guns from capture by advancing Confederates. During the heat of combat, 40 of the battery's 100 soldiers were killed or wounded. One gunner had failed to clear a weapon before it was fired and was nearly crushed to death by the stout recoil.

Gibbon, an acknowledged expert on artillery operations, had written the *Artillerist's Manual*, which was being used by the army at the time. He noticed that one of the battery's six guns was

Union soldiers, killed in action at the Battle of Gettysburg. Artillery, often firing canister against advancing enemy infantry, took a fearful toll during the Civil War.

rapidly firing double canister at the onrushing enemy, but that its angle of fire was well high of its mark. Jumping from his horse, Gibbon personally manipulated the elevating screw to adjust the angle and shouted, 'Give 'em hell, boys!' The ensuing shot blasted away a section of rail fence. The bodies of Confederate soldiers were torn to shreds, and one Union officer remembered seeing a severed arm reach a height of 9m (30ft) in the air.

With all six guns firing double canister, Battery B repulsed three Confederate charges in a span of only 20 minutes. With its ammunition nearly gone and 26 of its horses dead, Battery B remained in dire straits. Gibbon, however, managed to gather enough draft animals to withdraw in good order.

Doomed charge at Stone's River

To the west, during the Battle of Stones River, in Middle Tennessee, one of the most memorable casualties of the fighting was Union Colonel Julius P. Garesché. Serving as chief of staff to General William S. Rosecrans, commander of the Army of the Tennessee, Garesché was riding at his commander's side on 31 December 1862. As Rosecrans ignored Confederate artillery fire and galloped down a slope towards a faltering infantry line, another officer noticed a shell narrowly miss the leader and strike Garesché squarely in the face. The colonel's horse continued for 20 paces before the headless body, spouting blood profusely, fell to the ground.

Two days later, the Battle of Stones River ended when a Confederate charge, which had moments earlier broken a Union defensive line, was carried forward without orders, across the river and 549m (600yds) of open ground, directly into the teeth of 58 Union guns under the command of Major John Mendenhall and massed along commanding high ground. The cannon tore the attackers apart, firing at a rate of 100 rounds per minute. One Union soldier described the slaughter as if the Confederates had mistakenly 'opened the door of

During the Civil War, an observation balloon designed by Thaddeus Lowe floats aloft. Lowe used balloons to relay information about enemy troops and to spot artillery.

The high water mark of the Confederacy, Pickett's Charge at Gettysburg. Confederate troops are raked by Union cannon and rifle fire while crossing open ground.

Hell, and the devil himself was there to greet them.' In the Confederate 18th Tennessee Infantry Regiment alone, six colour bearers were struck down during the ferocious bombardment.

In addition to the engagements at Antietam and Stones River, concentrated artillery fire was decisive in numerous other battles of the Civil War. Confederate artillery slaughtered attacking Union troops at Fredericksburg, while Union artillery tore great gaps in the Confederate ranks at Malvern Hill during the Seven Days' Battles and doomed Pickett's Charge at Gettysburg.

The phenomenal industrial capacity of the factories in Northern states is illustrated by the fact that the Union army possessed roughly 2300 artillery pieces, including field artillery and coastal weapons, at the beginning of the Civil War. By war's end, the artillery had grown to more than 3300 cannon, over half of which belonged to the field artillery. It was further reported that the army had received just under 8000 guns, nearly 6.5 million projectiles, 2.9 million rounds of fixed ammunition, 45,000 tons of lead, and 13,300 tons of gunpowder.

Weapons of renown
Among the exceptional artillery pieces of the Civil War were massive 13in seacoast mortars deployed by the Union forces. Weighing 7.7 tonnes (8.5 tons), these monstrous weapons were capable of hurling a 100kg (220lb) shell 3932m (12,900ft). The most famous of these mortars, the Dictator, was mounted on a reinforced railroad car and

served for approximately three months with the 1st Connecticut Heavy Artillery during the siege of Petersburg, Virginia, in 1864.

A Confederate 18-pounder rifle, nicknamed 'Whistling Dick', was fired in defence of the Mississippi River city of Vicksburg, Mississippi during the siege of 1863. The 18-pounder, which had been converted from a Model 1839 smoothbore cannon, was actually credited with sinking the Union gunboat *Cincinnati*. After the fall of Vicksburg, 'Whistling Dick' disappeared, and the fate of the famous gun remains unknown.

The huge siege mortar known as the Dictator was transported atop a reinforced railroad car. The weapon could hurl a projectile nearly 4000m (13,123ft).

The most famous of Robert Parker Parrott's cannon was an 8in 200-pounder, which served in a battery located in a swampy area near Morris Island during the siege of Charleston. For two days during the summer of 1863, the gun, which was named the 'Swamp Angel' by its crew, battered the city. With the discharge of its thirty-sixth round, however, the gun exploded. Undoubtedly, this is one of the incidents to which the excerpt of Parrott's letter alluded.

The Franco-Prussian War
In the summer of 1870, strained relations between France and Prussia had deteriorated and both countries mobilized for war. From the beginning, the seven-month Franco-Prussian War proved

The 200-pounder cannon named the 'Swamp Angel' was designed by Robert Parker Parrott and deployed during the siege of Charleston, South Carolina, in the Civil War.

Although the French had developed these modern arms, they ultimately failed to use them to their fullest advantage. Nevertheless, on several occasions, the combination of the Chassepot and the Mitrailleuse took a heavy toll in Prussian casualties. The greatest deficiency in the French arsenal was its outdated artillery, which consisted almost entirely of antiquated muzzle-loaders of bronze construction, incapable of ranging effectively to duel with more modern Prussian guns. Prussian victories in the field were indeed won most often due to the preponderance of well-handled, rifled, breech-loading artillery, which literally pounded the French into submission.

Author Martin Tomczak notes that as early as the 1850s the Prussians were discussing the merits of rifled breech-loaders and actually experimented with prototypes constructed of steel barrels. A champion of change, taking on those in conservative military circles who resisted, was Crown Prince Wilhelm, who later ascended to the throne of a unified Germany. In 1859, the Crown Prince personally wrote the letter that trebled an order for 100 of the newer weapons to 300. An awareness that improved technology in small arms necessitated greater range for artillery contributed to this modernization effort. Simply put, it was suicidal to deploy gunners to service antiquated artillery pieces within range of the standard-issue rifles of enemy infantry. Having seen the success of modern artillery utilized during the American Civil War, the Prussians had fully re-equipped their army with rifled breech-loaders by 1867.

emphatically that technological advances in weaponry would vastly change the way wars were fought in the future. The French had introduced a new infantry rifle, known as the Chassepot, which was much superior to the shoulder arms of the Prussians and allied German states. Further, the French had developed the Mitrailleuse Model 1866, a 13mm weapon, which may be loosely classified as field artillery but is probably more accurately described as a forerunner of the modern machine gun. The Mitrailleuse had been invented by a French artillery officer and was composed of 25 barrels, which were fired in five tiers and encased by a bronze jacket. The weapon weighed roughly 798kg (1760lb) and required a team of six horses for transport. Its rate of fire was 130 rounds per minute, and its maximum range was 544m (1785ft).

Prussian Crown Prince Wilhelm, shown exhorting his troops during the height of battle, was a proponent of equipping the Prussian Army with modern, breech-loading cannon.

Within the next three years, the armies of all allied German states, including Saxony, Baden, Wurttemberg, Hessen-Darmstadt and Bavaria, were upgunned. Krupp, the dynastic producer of German arms, led the way in the development of steel capable of withstanding the rigours of combat without failure. By contrast, French arms manufacturers and steel producers during the 1870s were still struggling with the development of quality steel.

As Tomczak explains: 'The development by Krupp of steel strong enough to be used for gun barrels, coupled with the development of machine tools after mid-century, permitted the production of large numbers of guns of identical performance with easily replaceable parts. The strong barrels permitted the use of more powerful charges and thus enabled firing at longer ranges than before. All this assisted the Prussians in the development of a highly effective artillery arm. As a matter of interest a major reason for the lack of development of steel-barrelled guns in France was the fact that French industry was some way behind the Germans in steel-making and had not yet developed a strong-enough type of steel.'

Two weapons were predominant in the consolidated artillery arm of the German States at the beginning of the Franco-Prussian War. The 4-pounder and the slightly larger 6-pounder gun were used in light, highly mobile and heavier driving situations respectively. The 4-pounder, or 77mm, was accepted in 1864 as the Model C64 and later improved with a new breech block in 1867,

resulting in the designation C/64/67. The 91.6mm 6-pounder was of 1861 vintage, designated C/61, and when a new barrel was introduced in 1864 and found disappointing, the original version was reinstated.

While the weapons are described based upon the weight of the projectile fired, the names are somewhat misleading. The round shot fired by the 4-pounder actually weighed about 4.3kg (9.5lb),

and the round shot fired by the 6-pounder was about 6.9kg (15.2lb). Very similar in range, the 4-pounder maximum was approximately 3450m (11,320ft), while the 6-pounder could reach targets roughly 3440m (11,286ft) distant. Both weapons used bagged charges to propel their rounds.

Ammunition consisted of a variety of standard explosive and incendiary shells, along with the anti-personnel canister and shrapnel shells similar to those used to produce startlingly high casualties in the American Civil War. Percussion fuses detonated the shells on impact with the target. Prussian canister cylinders contained smaller shot composed of zinc, which was thought to exert less wear on the cannon barrel. The 4-pounder canister shot contained 48 small balls, and the 6-pounder round held 41 somewhat larger balls. Of particular interest are the incendiary shells, which, according to Tomczak, held about 20 per cent less explosive than normal shells. This was replaced by igniting elements called 'brander', which burned 15 to 20 seconds after impact and proved to be successful in setting structures on fire.

At the Battle of Spicheren on 6 August 1870, Prussian artillery pounded French position from a distance at which the outdated French weapons could not effectively reply. Twelve days later, at St Privat, the Prussians lost 8000 casualties, but once again their artillery tipped the balance of the fighting in their favour, eventually forcing the French army to fall back into the temporary safety of the fortress city of Metz.

During the decisive Battle of Sedan, fought at the end of August 1870, Prussian artillery, including 1000-pounder cannon with an innovative, horizontal sliding breechblock

THE FRENCH RESPONSE

The lesson of the importance of artillery in the outcome of the Franco-Prussian War was not lost on the French. In the decades following the humiliating defeat, a concerted effort was mounted to improve the performance of French artillery, culminating with the development of the famous 75mm field cannon.

Prussian artillery inflicted heavy casualties on the French, whose outdated cannon were unable to return fire effectively at the Battle of Spicheren.

mechanism, rained heavy shells upon the encircled enemy. One French officer realized that the heavy firepower had doomed his army and commented, 'We are in a chamber pot, and we are about to be covered in excrement.'

Initial negotiations for a French capitulation made little progress. However, the Prussian commander, General Helmuth von Moltke, warned that unless the French made a satisfactory response to demands for surrender his artillery would commence firing once more the following morning. Apparently, the threat of more shellfire was enough to cause the French to relent.

The Russo-Japanese War

The Russo-Japanese War of 1904–05 was significant for two reasons. Firstly, Japan, an Asian nation, actually confronted and defeated a traditional European power in Czarist Russia. Secondly, it was the first major war initiated during the twentieth century, and the world watched the power of concentrated heavy artillery

During the Battle of Sedan, heavy Prussian artillery pounded French positions into submission. Artillery was a decisive weapon throughout the Franco-Prussian War.

bring a defending garrison ensconced in strong fortifications to its knees. Located at the tip of the Liaotung Peninsula, the city of Port Arthur was the principal anchorage for the Russian Pacific Fleet.

The Japanese were victorious in a surprise naval action at Port Arthur in February 1904, blockading the surviving Russian ships inside the harbour. Subsequently, a Japanese army numbering nearly 90,000 soldiers landed on the peninsula and marched rapidly to lay siege to Port Arthur, which was defended by a contingent of 47,000 Russian troops manning 22 forts constructed in a wide arc along the hills to the landward side of the city. The

Russians possessed more than 500 guns, including those aboard the blockaded warships.

On 7 August 1904, a pair of Japanese 4.7in cannon thundered the opening shots of the battle for Port Arthur. The heavily defended Orphan Hills were captured by infantry charge following a 15-hour bombardment, but the victory cost the Japanese 1280 casualties. When the army commanders voiced their displeasure with the naval commanders over the fact that the Russians were able to call upon fire support from their warships in the harbour, the navy brought up a battery of 12-pounder cannon capable of suppressing the enemy fire.

Five days later, Japanese troops moved against the 174-Metre Hill. Their commander, General Maresuke Nogi, was reported to have been somewhat startled at the inability of the Russians to coordinate their powerful artillery and as a result he ordered a frontal assault. The tenacious Russian defenders were driven from the high ground, but the fighting cost the Japanese dear – another 1800 dead.

When it became apparent that the price of penetrating the defensive cordon would be too high, the two sides became locked in a protracted siege. The tide began to turn in favour of the Japanese with the arrival of more than 20 massive 11in siege howitzers, manufactured in Germany by Krupp. These giant weapons were capable of firing a shell weighing just over 227kg (500lb) a distance of nearly 9007m (29,550ft). The Japanese already had 450 cannon of various calibres on the field at Port Arthur, but the addition of the 11in howitzers gave them the heavy artillery and plunging fire that could pulverize the Russian fortifications. Eventually, the gigantic Krupp howitzers would fire more than 35,000 shells during the siege of Port Arthur, and the projectiles earned the nickname 'roaring trains' from the Russian troops.

Prussian General Helmuth von Moltke warned the French at Sedan to surrender or endure a relentless bombardment from the massed firepower of the Prussian artillery.

During a week of bitter fighting from 28 November to 5 December 1904, the Japanese assaulted the key positions on the western side of the Russian defensive perimeter, the 203-Metre Hill and another prominence named Akasakayama. The 11in siege howitzers fired more than 1000 huge shells at the Russians in a single day; nevertheless, the attackers lost more than 8000 men on the last day of fighting for the two hills.

When they reached the summit of the 203-Metre Hill, the Japanese brought up their 11in howitzers along with their huge armour-piercing shells and ravaged the Russian warships in the harbour at Port Arthur. The land-based artillery sank the battleship *Poltava* on 5 December, and the battleship *Retvizan* two days later. On 9 December, the battleships Persvyet and Pobieda were sent to the bottom of the shallow anchorage along with the cruisers *Bayan* and *Palada*. A fifth battleship, the *Sevastopol*, took hits from five 11in shells but refused to sink. After repeated torpedo attacks by Japanese destroyers, two of which the *Sevastopol*'s gunners sank, the warship was scuttled on 2 January 1905. On the same day, the Russians capitulated.

In the wake of the Russo-Japanese War, Captain A. Degtyarev, a Russian artillery officer, wrote an exhaustive analysis of the performance of artillery during the brief but destructive conflict. Both sides had realized notable successes, and the author speculated that the refinement of tactics would lead to even more impressive performance in the wars of the future:

'To be the first to open fire is a great step for success, more particularly when firing from a concealed position. If the enemy, notwithstanding this, succeeds in locating the battery, and there is danger of heavy loss to personnel, it is better to wait. In this case the enemy will soon change to slow fire, when the battery should again open with rapid fire for a short time. If the shells burst accurately, the opponents will probably cease fire. It was in this manner that the 3rd Battery of the 5th East Siberian Artillery Brigade engaged 18 guns, firing from three directions, for four days (27th February to 2nd March) at Kun-tu-li-chun. A battery of the 35th Artillery Brigade was equally successful in the fighting on the 12th of October, during which it lost only one officer and six men wounded. Such is the character of modern artillery duels, and the chief task of the artillery is not to destroy the hostile guns, but merely to threaten them, so as to prevent them from firing upon its own troops.'

Concealed positions

The Russian officer also advocated firing from concealment with fewer batteries concentrated in close proximity to one another. He had apparently reached this conclusion due to the nature of modern quick-firing artillery pieces:

'A battery skillfully posted in a concealed position may be able to silence several hostile batteries, thus leaving other batteries free to engage the enemy's troops. The 1st and 2nd batteries of the 35th Artillery Brigade carried out this duty admirably at the village of Khan-chen-pu on the 14th October. Being most skillfully posted, they silenced the Japanese artillery at Scha Ho Station, thus making it possible for the 4th and 5th batteries of the same brigade to engage the hostile infantry concentrated at the village of Sha-ho-pu. A well-concealed position is the best security against defeat.'

Two positions of relative concealment offered the best advantage, according to Degtyarev. These were the rear slope of a hill or the rear of a hill with two other hills in front of it with the valleys separating them at least 400 to 500 paces wide. Such natural defences would make counter-battery fire ineffective, with most enemy shells passing overhead or detonating against the mass of the protecting hill:

'At Kun-tu-li-chun three batteries of the 5th East Siberian Artillery Brigade occupied a position similar to [the rear slope of a hill] and successfully engaged 18 hostile guns from the 27th February to the 2nd March, during which time they only lost one officer and 18 men. Occasionally for 20 minutes or so the air above the battery seemed full of bursting shells, but the elevation was too high. At such times the battery would remain silent, its gun detachments taking shelter in lightly protected trenches.'

As they were to prove 40 years later during World War II, the Japanese had already shown an adaptability to night combat during their war with the Russians. For this reason, apparently, Degtyarev chose to discuss the defensive employment of artillery engaged in night operations.

'The cooperation of the artillery is an important factor in the defence by night. If skillfully used, it will not only have great moral(e) effect, but will also inflict serious losses on the enemy... Fire directed down valleys perpendicular to the front of the position is particularly useful because such valleys are natural lines of advance for the attack. In this connection, observation to ensure the correction of fire is most important. The best plan is to post observers in the firing line, and to connect them with the batteries by telephone. To ensure the

Russian artillerymen ready their weapon during the defence of Port Arthur. Both sides used heavy artillery liberally during the Russo-Japanese War.

accuracy of the fire directed on the ground in front, the ranges and bearings to important points in front of the position should be ascertained by daylight, especially if such points afford facilities for the concentration of hostile infantry. Successful correction of the elevation and direction will remove all danger to one's own troops.

'The 2nd and 3rd batteries of the 5th East Siberian Artillery Brigade were brilliantly handled during the night fighting at Kun-tu-li-chun between 2nd and 7th March when the batteries were posted 1500 paces behind the infantry … the front of the position held by the 5th East Siberian Rifle Division was intersected at right angles by two deep valleys, the mouths of which, at a distance of some 700 paces from the front of the position, were closed by two villages. These villages offered the only available cover for the concentration of the enemy's infantry, but the Japanese were successfully driven out of them by the action of the artillery alone.

The shape of things to come

References to quick-firing guns, telephones and forward observers indicate that by the dawn of the twentieth century technology had begun to usher in a new age for the role of field artillery. Observers from European countries and from the United States were present during the Russo-Japanese War and probably reported to their respective governments on the destructive capability of unleashed modern artillery. The killing

Assailed by Japanese artillery positioned in the surrounding hills, Russian warships are pulverized at Port Arthur.

fields of World War I would further validate their case and provide an additional revelation – this one more horrific than could have been imagined.

By 1890, all of the major armies of the West had incorporated rifled, breech-loading artillery into their organic orders of battle, but the elder statesmen of European and American military hierarchies were slow to modify infantry tactics to deal with the enhanced capabilities of artillery, the machine gun and, later, the tank and the aircraft. In consequence, thousands of men met their deaths by executing orders based on tactical maxims that predated the Napoleonic Era. As in every war before or since, the general issues the orders, and the common soldier is required to pay the substantial price for any miscalculation.

WORLD WAR I

The conflagration of World War I, precipitated by an assassin's bullet in the Balkan city of Sarajevo, elevated the concept of modern, total war to a level beyond imagination. Political tensions, imperialism driven by economic necessity and nationalistic fervour, and an ever-escalating arms race resulted in the eruption of hostilities in the summer of 1914.

A prolonged war became inevitable, and armies of unprecedented size suffered staggering casualties within minutes, attacking opposing trenchlines across a barren, cratered moonscape that came to be known as No Man's Land. On the Western Front, the static trenches stretched from the North Sea to the Swiss frontier, and the refinement of the technology of killing reached its zenith.

The rapid-firing machine gun mowed down lines of infantry, whose commanders had, more often than not, failed to come to grips with the idea that improved weaponry inevitably necessitates a change in tactics. The artillery of the warring armies had, for the most part, been steadily improved and upgraded during the previous 60 years; however, on the eve of World War I its incredible destructive potential had not yet been fully realized. Millions

Left: The massive wheeled carriage on this heavy British weapon is plainly visible. During an exercise, the gun's crew prepares to load and fire while observers look on.

of dead and wounded would eventually bear out the premise that more guns of larger calibre and greater accuracy could have a telling effect.

During the course of World War I, the evolving conditions of the battlefield gave rise to tactics that were better suited to prevailing conditions. In 1914, an artillery barrage was general in nature, intended to inflict casualties upon the enemy on a broad front when preceding an infantry assault. However, the depth of trenches and bunkers, the killing power of machine guns and concentrated rifle fire, and the disappointing performance of some artillery shells that failed to deliver decisive blows, meant that the combatants were required to reshape their offensive methodology.

The British, for example, began to employ shorter bombardments with targets spotted by aerial reconnaissance. Instantaneous fuses exploded on impact with the ground rather than expending much of their explosive capability after penetrating a few inches. Therefore, they were more effective anti-personnel weapons and better able to clear

obstacles such as barbed wire and machine-gun positions. The general barrage was replaced with a systematic approach. The creeping barrage and box barrage might neutralize a certain position or force the enemy to maintain cover while an infantry operation took place nearby. The use of smoke or gas shells could have the same effect, keeping enemy troops immobilized or concealing one's own movements.

The British 18-pounder field gun was often used in the creeping barrage to provide a curtain of fire roughly 46m (50yds) ahead of advancing infantry, suppressing return fire from German trenches. A field manual on the topic concluded:

'The barrage does not lift direct from one trench to another, but creeps slowly forward, sweeping all the intervening ground in order to deal with any machine guns or riflemen pushed out into shell holes in front of or behind the trenches. This creeping barrage will dwell for a certain time on each definite trench line to be assaulted. The infantry must be trained to follow close behind the barrage from the instance it commences and then, taking advantage of this "dwell", to work up as close as possible to the objective ready to rush it the moment that the barrage lifts.'

In concert with the creeping barrage, larger guns such as the 4.5in howitzer, would often conduct an in-depth barrage, perhaps targeting lines of communication, while heavier 60-pounders assaulted lines of advance. The heaviest calibre weapons bombarded areas where enemy troops might congregate for a counterattack, or enemy artillery positions that could potentially ravage the ranks of the advancing infantry.

Improvements in target acquisition developed along with the heavier guns and revised tactics. Two primary methods of locating enemy gun emplacements were sound-ranging and flash-spotting. Sound-ranging involved the use of microphones to detect sound waves generated by the firing of a distant weapon. Measuring the time interval between a series of these, microphones placed at certain locations led to the plotting of several coordinates on a map, fixing the location of the gunfire. Flash-spotting remained in use until the end of World War II and involved observers noting the azimuth of the fire – the number of degrees clockwise from due north – in order to fix the source.

The French 75

Following their humiliating defeat in the Franco-Prussian War, senior French commanders embarked on a programme of modernization of their nation's armed forces. Their effort included not only improving weaponry, but also the re-establishing of the esprit de corps and the honour of the nation's armed forces. The painful lessons learned while opposing the capable and efficient Prussian artillery resulted in a programme of research and development, which was to pay a great dividend.

The introduction of the French 75mm Model 1897 field gun was a watershed development in the history of modern artillery. Known as the French 75, the Soixante Quinze, and sometimes simply as the 75, this weapon combined a self-contained recoil system, modern sighting, fixed shell ammunition, a protective shield for the artillerymen servicing the gun and a fast-action breech mechanism. In short, the quick-firing French 75 embodied the combination of decades of advances in artillery technology.

The 75mm was the culmination of a design effort that had begun with a 57mm weapon that had undergone field trials in 1891, and was developed at the request of officers who had seen the potential for success in a somewhat heavier cannon. More than 21,000 French 75s were produced during a remarkably lengthy period of manufacture, stretching from 1897 to approximately 1940. Initially, a team of six horses was required to haul the gun, which weighed just over 1180kg (2600lb). Its rate of fire was an impressive 15 rounds per minute at a maximum range of 6840m (7480yds). An anti-aircraft version

The British 18-pounder field cannon, shown here being serviced in a camouflaged position along a dirt road, was recognizable by its protective gun shield.

155MM RIMALHO

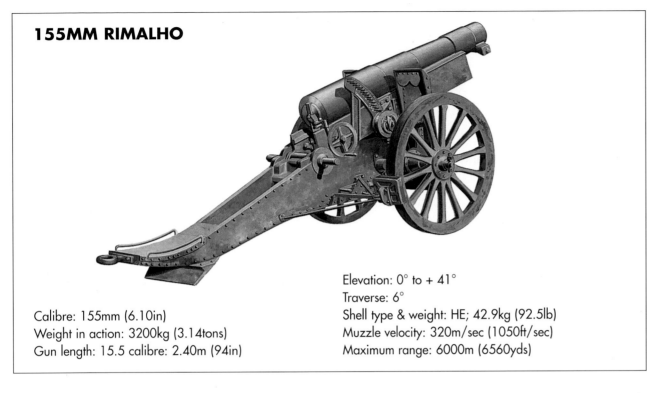

Calibre: 155mm (6.10in)
Weight in action: 3200kg (3.14tons)
Gun length: 15.5 calibre: 2.40m (94in)

Elevation: 0° to + 41°
Traverse: 6°
Shell type & weight: HE; 42.9kg (92.5lb)
Muzzle velocity: 320m/sec (1050ft/sec)
Maximum range: 6000m (6560yds)

The French 75mm Model 1897 achieved lasting fame during the crucial battles of the Marne in the summer of 1914 and at Verdun during a protracted eight-month struggle from February to September 1916. At Verdun alone, the weapons were reported to have fired an astounding 16 million rounds, accounting for nearly 70 per cent of all the shots fired by French artillery during the battle. More than three million shells were fired in a three-day period during the French offensive in the Verdun vicinity the following spring. The 75 also became the primary deliverer of the phosgene and mustard-gas shells fired by the French, particularly during the static trench warfare, which developed as the fighting ground to a stalemate.

When the United States entered the war in the spring of 1917, its army was ill-prepared for battle. Nearly 500 batteries, each fielding four guns, were equipped with the French 75, and American industry began to produce the gun in its own right under licensing. By the end of the war, nearly 1100

of the gun, mounted on a truck, was adopted by the army in 1913.

The French 75 was produced by government manufacturers, such as the Atelier de Construction de Puteaux. It commonly fired two types of ammunition early in the war, a fragmentation or shrapnel shell weighing 5.3kg (11.7lb) and a high explosive shell of nearly 7.25kg (16lb). The principal advantage of the 75 was its impressive rate of fire due to the improved 'hydro-pneumatic' recoil system, which relieved the crew of having to roll the weapon back into place and reacquire the target following each round.

Superbly designed for use against massed troops in the open, the 75mm gun was indeed a field weapon. When war broke out in August 1914, more than 4000 of the guns were already in service with the French Army. During the course of the war, more than 200 million shells were produced by private armaments manufacturers, and by 1915, the insatiable appetite for the ammunition had dictated a 500 per cent increase in daily production, from 20,000 to 100,000 rounds. In order to achieve such numbers, the French government gave contracts to civilian companies, and the result was often an inferior-quality shell.

A French 75mm field cannon is hauled to its firing position on a horse-drawn limber. The 75mm cannon was still in service during World War II.

HEAD TO HEAD: *Canon de 75 mle 1897* VERSUS

Perhaps the most famous and successful field-artillery piece of World War I, the French 75mm cannon became a mainstay of the Allied forces on the Western Front. The versatile weapon, also known as the Soixante Quinze, was still in use during World War II.

Canon de 75 mle 1897

Calibre: 75mm (2.95in)
Weight: travelling 1970kg (4343lb);
    in action 1140kg (2514lb)
Gun length: 2.72m (8.92ft)
Elevation: -11° to +18°
Traverse: 6°
Shell type & weight: 6.2kg (13.66lb)
Muzzle velocity: 575m/sec (1886ft/sec)
Maximum range: 11,110m (12,140yds)

STRENGTHS

• Self-contained recoil system
• High rate of fire
• Modern sighting

WEAKNESSES

• Ineffective against fixed fortifications
• Light calibre
• Lack of mobility

77mm Field Gun M96nA

The most common field cannon in the German arsenal at the beginning of World War I, the original 77mm gun was outclassed by its French 75mm counterpart, resulting in a major revision. Its distinctive report earned the weapon the nickname Whizz Bang from Allied soldiers.

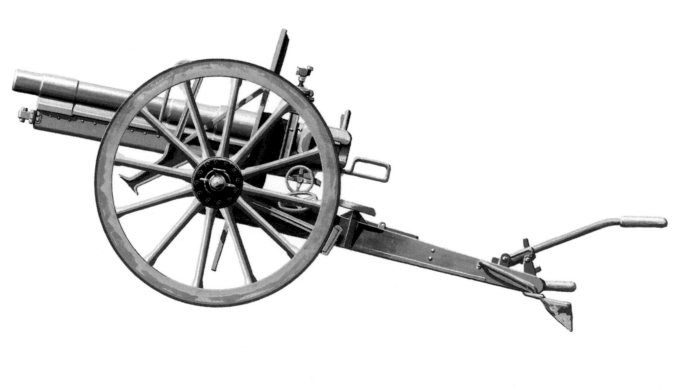

77mm Field Gun M96nA

Calibre: 77mm (3.03in)
Weight: 925kg (2039lb)
Gun length: 27.3 calibre: 2.1m (82.67in)
Elevation: -13° to +15°
Traverse: 8°
Shell type & weight: shrapnel; 6.85kg (15.1lb)
Muzzle velocity: 465m/sec (1525ft/sec)
Maximum range: 7800m (8530yds)

STRENGTHS

- Available in large numbers
- Sliding wedge breech
- Good mobility

WEAKNESSES

- Low rate of fire
- Lightweight shells
- Used in static role

had been produced in the United States. About 140 of these actually reached the battlefields of France. Harry S. Truman, the future president of the United States, served as a captain of 75mm guns attached to Battery D, 129th Field Artillery, at the Meuse-Argonne in the autumn of 1918.

For all of the zeal in the modernization of the French armed forces, the dependence on the 75mm gun by senior commanders resulted in one miscalculation, which cost the lives of thousands of French soldiers during the Great War. As effective as the weapon was during mobile war and in support of troops on the move, it lacked the firepower to inflict serious damage on fixed fortifications or well-prepared trench works. The number of 75s far exceeded those of higher calibre available to the army until 1917, when sufficient numbers of heavier guns such as the 155mm Schneider howitzer began to arrive.

The Schneider was capable of firing a 43kg (95lb) shell up to 10,973m (12,000yds). Compared to the 75, however, its rate of fire was ponderously slow, at only two rounds per minute. The French were also able to eventually deploy the 58T trench mortar, which fired a 15.8kg (35lb) shell packed with just over 5.9kg (13lb) of explosives a distance of 457m (500yds), at an elevation of 45–80°. Another 75mm weapon, the Schneider Model 1912, was developed primarily for use with cavalry units.

Infantryman Henri Barbusse of the nocturnal bombardment of Artois in 1915 recalled:

'The rumbling of the artillery became more and more frequent and ended up forming a single rumbling of the whole earth. From all sides, outgoing bursts and explosions threw forth their flashing beams which lit up the dark sky over our heads with strips of light in all directions. Then the bombing grew so heavy that the flashes became continuous. In the midst of the uninterrupted chain of thunder claps we could see each other directly, helmets streaming like the bodies of fish, gleaming black iron shovels, and the whitish drops of the endless rain, truly it was like moonlight created by cannon fire.'

In 1914, the most numerous heavy weapon in the French arsenal was the Rimailho 155mm howitzer, which was roughly equal in stature to the 5.9in howitzer fielded by the Germans. The Rimailho 155 was also the most modern of the French heavy artillery, and its capabilities were sorely needed, as commanders came to the realization that heavier weapons with relatively high elevation to create plunging fire were most effective on hard targets. Author Charles Messenger explains:

The American crew of a French 75mm field cannon loads and fires its weapon with efficiency. A pile of spent shell casings is testimony to the day's work.

'It had a particularly high angle of fire, and was thus very useful in trench warfare, where the more steep the fire of howitzers and trench mortars, the more valuable they were. This was even truer of heavy calibre weapons than of light calibre ones, as the former were effective against personnel and the entrenchments themselves, the latter against personnel, and only to a small extent against fortifications.'

Another observer wrote:

'French gunners were devoted to the 75 as they quite rightly expected it to devastate enemy formations in the open. However, it was not intended, nor consequently suited, for high angle indirect fire of the sort needed to deal with an entrenched enemy, and its shells were also too light to be effective against such positions in any event. Yet these capabilities were not considered to be important in the years leading up to World War I and to cater for them with slow heavy guns would impede the advance and freedom of action of French units in the field. Indeed, the low ratio of heavy guns was seen as a powerful symbol of offensive intent and became a source of martial pride. In 1909 the general staff representative on the Chamber of Deputies' Budget Commission was heard to declare: "You talk to us of heavy artillery. Thank God we have none. The strength of the French Army is in the lightness of its guns." Thus, by 1914, approximately 3800 75s were held by the French Army as opposed to a mere 389 heavy guns, most of which were obsolete fortress and siege designs.'

Future US president Harry Truman served as a captain in the US field artillery during World War I. Truman commanded a section of French 75s.

The 155mm Rimalho cannon is shown in the protective position of an earthen revetment. Heavier weapons proved more effective.

By the time German pressure on the beleaguered French fortifications at Verdun had begun to ease in the late summer of 1916, General Charles Mangin was not content to simply lick his substantial wounds. He voiced to superiors and to the newspapers his intent to strike hard at the enemy.

'I box in the first line with 75s. Nothing can pass through the barrage; then we pound the trench with 155s and 58s... When the trench is well turned over, off we go. Generally, they come out in groups and surrender. While this is going on, their reserve companies are pinned in their dugouts by a solid stopper of heavy shells. Our infantry waves are preceded by a barrage of 75s; the 155s help bang down the cork on the reserve companies; the tides of steel join up with the infantry 70 or 80 yards behind. The Boche gives up... You see, it is all very simple.'

Simple though the concept of tactical victory may have been, the fact remained that the German enemy had prepared for war as well. Artillery, both light and heavy, had been a key component of that planning. The French were by no means immune from the destructive power of the enemy guns.

The British Expeditionary Force

The experience of the British Army during the Boer War of 1899–1902 prompted an evaluation of artillery performance and the adoption of more modern weaponry, such as the 13-pounder quick-firing gun adopted by the Royal Horse Artillery and the 18-pounder placed in service with the Royal Field Artillery. Both of these weapons entered into general service in 1904 and remained mainstays of the artillery throughout World War I.

The 18-pounder was a workhorse of the British Army artillery, and more than 1200 of the guns were in service at the outbreak of the war. The number of these weapons in service increased to nearly 9500 by November 1918. The earliest version of the 18-pounder was capable of hurling an 18lb 8oz (8.4kg) shell over 5944m (6500yds). An improved series, the Mark II, which hastened the rate of fire with an upgraded recoil function, was introduced in 1916, and a further refinement of the weapon, designated the Mark IV, with a more efficient loading apparatus, was introduced late in the war. Successively, the maximum elevation of the gun increased from 16° to 30°, and its range was extended to 8504m (9300yds). The Mark IV rate of fire was an impressive 30 rounds per minute.

The 13-pounder, a 76mm weapon, fired a projectile that actually weighed 5.7kg (12.6lb) at a range of 5395m (5900yds). During the retreat of the British Expeditionary Force from Mons and the Battle of Le Cateau in the summer of 1914, the 13-pounder was used extensively. However, as mobile warfare waned in the autumn of that year, the 13-pounder was found to be relatively ineffective against enemy trench works and gave way to the employment of the somewhat heavier 18-pounder.

As the aircraft emerged as a weapon of war, the 13-pounder found new usefulness as a medium anti-aircraft gun. In this role, the weapon served to bridge the inter-war years and was active with the British Army essentially until its evacuation from the continent at Dunkirk in 1940. It was then replaced with a 3.7in anti-aircraft weapon.

The 4.5in howitzer entered service with the Royal Field Artillery in 1909 and gradually replaced an outdated 5in howitzer, which saw a great deal of action during the first two years of the war. Its shell, weighing 15.8kg (35lb), was the equivalent of a 114mm (4.5in) round, capable of reaching targets 6675m (7300yds) distant at an elevation of up to 45°. The sliding block breech was

A Veritable Cloud

Massed artillery fire was a primary contributor to the proverbial fog of war and the rattle of battle, often producing huge clouds of acrid smoke and raising such a din that artillerists suffered permanent hearing loss. Soldiers remembered the experience of a sustained artillery bombardment as horrifying, giving rise to the term 'shell-shocked'.

redesigned and introduced with the Mark II version of the gun in 1917. The remarkable service life of the 4.5in howitzer with the British Army extended until the end of World War II.

At the beginning of World War I, the 4.7in quick-firing field gun, which had come into service 30 years earlier and was a veteran of the Boer War, was widely used. Its 20.8kg (46lb) projectile could travel 9144m (10,000yds), and 16 of them were used at the First Battle of Ypres. However, limited supplies of ammunition often meant that only eight rounds per day were allotted to fire-support assignments. The gun was unwieldy and difficult to position on the battlefield, eventually being largely replaced by the 6in howitzer. A pair of 15-pounder (84mm) guns were placed in service with the British Army, some of which had been manufactured in Germany. With the exception of a few batteries that remained in service in the Middle East, most of these were withdrawn by 1915 and used for training purposes.

Antiquated 8in and 15in howitzers, both redesigned versions of older naval and army weapons, were used early in the war and replaced by more modern weapons as soon as possible. The 8in howitzer had been rebored from a 6in cannon and fired a 90.7kg (200lb) shell up to 11,247m (12,300yds) before being replaced by the 4.5in and 9.2in howitzers. The 15in howitzer fired a large-calibre shell up to nearly 10,058m (11,000yds). By 1916, it was rapidly being taken out of service.

Desperate hours at Le Cateau

As the British withdrew from Mons, General Horace Smith-Dorrien, commander of II Corps, ordered his exhausted men to turn on the pursuing Germans and make a stand. Early on the morning

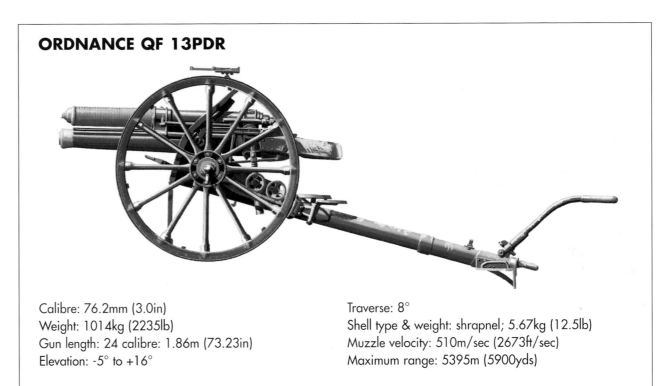

ORDNANCE QF 13PDR

Calibre: 76.2mm (3.0in)
Weight: 1014kg (2235lb)
Gun length: 24 calibre: 1.86m (73.23in)
Elevation: -5° to +16°

Traverse: 8°
Shell type & weight: shrapnel; 5.67kg (12.5lb)
Muzzle velocity: 510m/sec (2673ft/sec)
Maximum range: 5395m (5900yds)

of 26 August 1914, the Germans attacked. Their initial attempt to breach the British line was repulsed by artillery and small arms fire. Then, around 10 a.m., they tried again.

According to the regimental journal of the Lancashire Fusiliers:

'The German guns unleashed a maelstrom of steel inflicting heavy losses to the besieged British forces. The men of the King's Own Yorkshire Light Infantry, and the Suffolks who held the front line, were literally pulverized by high explosives and shrapnel. Now the Imperial German Army began their attack. Like a great grey sea the Germans advanced on a two-mile front from the valley of the Selle to Rambourlieux Farm. These huge formations proved deadly for the Germans. What followed can only be described as a massacre. The rifles and field guns of the BEF took the lives of many brave Germans. The toll was also great on the British. The unceasing artillery, machine gun and rifle fire flailed the men of the BEF. Only one gun of 11 Battery was now operational. The guns of the remaining batteries rained death on the advancing Germans. The guns of 122 Battery wiped out a whole platoon of Germans. The others broke for cover.'

During the fighting early in the war, massed troop formations were regularly subjected to withering combinations of fire from strong enemy

The British QF 4.7in field gun was a converted naval weapon that saw action during the Boer War and was widely used during the early years of World War I.

positions. The advantage in combat lay with an army that was fighting defensively from well-prepared entrenchments. Therefore, it must be acknowledged that the era of trench warfare, which began late in 1914 and extended four years, was precipitated by the modernization of weaponry, which inflicted terrible losses on exposed formations.

The brave men of II Corps fought tenaciously at Le Cateau, but they were outnumbered at least three to one, and the weight of the German attacks began to take their toll. By noon, the situation was tenuous at best:

'Machine gun fire and rifle fire was [sic] battering the Suffolks, and the final gun of 11 Battery was now out of action. Hastily the men of the 2nd Manchesters, the Argylls and Sutherland Highlanders were ordered up to help the beleaguered Suffolks but they were strafed by enemy fire and suffered severe losses. Steadily the

German attack continued. Towards the left flank the Germans mounted an attack but were repelled by the rifle fire of the BEF. Again the valiant Germans rallied and charged. Under a hail of bullets, some determined forces made it across the Cambrai Road but were chopped down by the machine guns of the Royal Scots near Audencourt. To the far left the British were under heavy attack from artillery and machine gun fire. The King's Own Royal Lancasters had 400 casualties.'

An hour later, the British were on the verge of being overwhelmed. Individual acts of heroism became commonplace:

The British Expeditionary Force retreats from Mons ahead of the German onslaught in the summer of 1914. Field artillery was used extensively to cover the withdrawal.

'The situation was now critical. The guns of 122, 123 and 124 Batteries were in jeopardy. A call went out for volunteers to retrieve the guns. Captain R. A. Jones of 122 Battery gathered his men and six fearless teams of horses and was ready to attempt the task. Jones inspired his men by leading them down the battered hillside towards the guns. German machine gun fire tore through the teams. Eight men, including Captain Jones, and 20 brave steeds fell. The other teams reached the guns and limbered up three of them. One team was shot down in the road. The other two galloped up the hill past the men of the West Kents. They waved their hats and cheered wildly as they galloped by. "It was a very fine sight," stated one British officer.

Left: General Sir Horace Smith Dorrien commanded the II Corps and the Second Army of the British Expeditionary Force during World War I.

Acts of courage filled the battlefield, but none were more daring than the story of Captain Reynolds of the 37th Battery. With a handful of volunteers they charged forward to rescue two howitzers that had been left in the open. With the Germans only 200 yards away the men limbered up the guns. They galloped away but one team fell to artillery fire. The other team of Reynolds and drivers Luke and Drain made it back. For their bravery in the face of death they received the Victoria Cross.'

The nature of combat in the twentieth century had, by the time of the Great War, fostered a re-evaluation of the role of heavy artillery. While much effort had been placed in the improvement of field artillery by the world's major armies, the development of heavy artillery for modern warfare had lagged behind in some military establishments. Heavy artillery had previously been considered a siege weapon – fixed, and used primarily for the reduction of enemy fortifications. While this was still the case, the introduction of better mobility might bring the heavier guns to bear against an adversary whose trench systems included deep

bunkers a number of feet below ground level and trenches with limited exposure, which were virtually impossible to destroy using lighter field artillery shells unless a direct hit was scored. By the spring of 1918, Allied artillery on the Western Front was firing between one million and three million rounds per week.

Dealing with the enemy trenches

The heavy artillery of the British Army was serviced by units of the Royal Garrison Artillery. During the fighting on the Western Front, it became readily apparent to the combatants that heavy artillery would be necessary to successfully reduce enemy fortifications. The British actually had gained some experience in this regard during the Boer War, and their 60-pounder gun was developed with that experience in mind and deployed in 1904.

Firing a 127mm (5in) shell a maximum distance of 9418m (10,300yds), the earliest version of the 60-pounder weighed nearly 4.5 tonnes (4.9 tons). Initially it was hauled into position by a team of 12 horses. However, a later version, weighing one ton

Stripped to the waist in the oppressive heat, Turkish artillerymen load and fire their weapon against ANZAC troops near Gallipoli.

more, was towed by a steam tractor. Wartime modifications improved the transportation of the weapon and increased its range to 11,247m (12,300yds). The 60-pounder proved to be one of the most popular heavy weapons in the British arsenal, and it remained in service until the end of World War II.

Several modern, large howitzers appeared on the battlefields of World War I in service of the Royal Garrison Artillery. Among the most widely used was the 6in howitzer, which, in its original configuration, fired a 53.75kg (118.5lb) shell a maximum distance of 4755m (5200yds) with a maximum elevation of 35° for effective plunging fire. Both the range and elevation were a bit of a disappointment initially. An improved model increased the elevation to 45° and more than doubled the weapon's effective firing distance to 10,424m (11,400yds). According to records, more than 3600 of the 6in howitzers were deployed during the war, and they fired more than 22 million rounds.

Weighing 15.5 tonnes (17 tons) and assembled in three sections, the 9.2in howitzer was designed for fixed use. The entire assembly process took 36 hours, and the weapon was transported on three specially built carriages. In order for the weapon to fire, a box containing 8 tonnes (9 tons) of earth was attached forward to prevent it from literally leaping off the ground with each shot. The Mark I went into production in July 1914 and was first used during the Battle of Neuve Chapelle in the spring of 1915. Its shell weighed 131.5kg (290lb) and

Churchill's Folly?

In his capacity as First Lord of the Admiralty, future British prime minister Winston Churchill was a staunch advocate of the landings of Commonwealth troops on Turkish shores at Gallipoli in 1915. The operation appeared doomed from the start, as concentrated Turkish artillery, machine-gun and rifle fire exacted horrendous casualties. Churchill received a great measure of the blame and was demoted as a result. He later resigned from a minor post in the British government and commanded the 6th Battalion of the Royal Scots Fusiliers on the Western Front. Nearly 30 years later, during World War II, Churchill again proposed an amphibious landing – this time at Anzio in Italy. The result was a lengthy stalemate.

could reach targets 9190m (10,050yds) distant. The Mark II improved the range to 12,742m (13,935yds) but required an additional 1.8 tonnes (2 tons) of earth in the forward stabilizing box. A total of 450 of the 9.2in howitzers were built between 1914 and 1918.

In the summer of 1916, the mammoth 12in howitzer was first seen on the battlefield. The Mark II weapon weighed 34 tonnes (37.5 tons), and its earthen box counterweight contained 18 tonnes (20 tons) of soil. The Mark I and Mark III were mounted on rail lines. The arduous task of loading the weapon required that the barrel be depressed almost to horizontal for the 340kg (750lb) shell to be inserted. The maximum range was 10,369m (11,340yds) at an elevation of 70°. The later Mark IV version employed a longer barrel, resulting in an increased range to 13,122m (14,350yds). The mere presence of the 12in howitzer provided a morale boost to

Commonwealth troops, and a direct hit from the powerful weapon was devastating.

Under the gun at Gallipoli

The disastrous 1915 Allied campaign in the Dardanelles provided stark evidence of the potential domination of a battlefield by artillery. In March, Turkish artillery contributed to the defeat of an attempt by battleships of the Royal Navy and the French Navy to silence coastal defences and force the strait open. The French battleship *Bouvet* sustained eight hits from Turkish guns, and was one of several warships to be damaged by the accurate fire. In possession of the high ground surrounding the landing areas at Gallipoli, Turkish artillerymen had ample opportunity to range their weapons.

Colonel Hobbs of the 1st Australian Division wrote in the spring of 1915:

'We are dominated by the enemy guns and observation stations, and we are subject to heavy shell fire at once. Our guns are badly knocked about and many casualties [sic]. It is impossible for us to discover enemy batteries which are hidden in nullahs and behind ridges which our guns cannot reach.'

Canadian artillery at Vimy Ridge

During three days in April, 1917, Canadian forces spearheaded an offensive against German positions on Vimy Ridge in France. Their victory is considered by many to be a defining moment in establishing a Canadian national identity. Harry

ORDNANCE BL 60PDR MK 1

Calibre: 127mm (5.0in)
Weight in action: 4470kg (9655lb)
Gun length: 33.6 calibre: 4.29m (14.06ft)
Elevation: -5° to +21.5°

Traverse: 8°
Shell type & weight: shrapnel; 27.22kg (60lb)
Muzzle velocity: 634m/sec (2080ft/sec)
Maximum range: 11.25m (12,300yds)

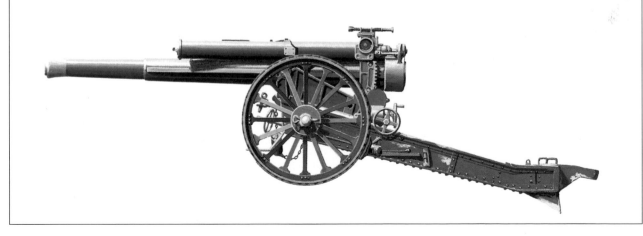

Whitfield Mollins of Moncton, New Brunswick, served in the artillery and recorded his routine before and during the attack in a diary, which he carried in the pocket of his tunic. On April 5, he wrote, 'Carrying ammunition to the No. 4 gun all day. We are getting a reserve store ready for the 'strafe' which comes off in a few days.'

Other entries from Mollins include:

'On duty at the guns all day. Did considerable firing… On ammunition fatigues all day. Did considerable firing. Had to take a count of all the shells et cetera… On the guns all night. Fired 123 rounds… This is the day of the big 'strafe'. There was heavy bombardment at 5:30 a.m. and the infantry went over from Vimy Ridge to the right. Large numbers of prisoners and guns were taken… On the guns all day… We started firing about 11.30 a.m. and from then till six fired 215 rounds.

Fritz shelled quite heavily today. While we were firing the Huns burst two shrapnels right over [our] gun. No one was hurt. Spent the night in an old cellar, which we cleaned out.'

Killing ground on the Somme

Perhaps no other episode during World War I illustrates the scale of sacrifice and suffering inherent in the bloody battles on the Western Front as the Somme during the summer and autumn of 1916. On the first day of their offensive, 1 July, the British lost no fewer than 57,470 casualties, 20,000 of these killed in action. It was the bloodiest day of World War I – and the bloodiest day in British military history. Intended as an offensive to relieve pressure on French troops who were under attack at Verdun, the initial British infantry assaults had been preceded by eight days of artillery bombardment, in an attempt to destroy German positions and barbed-wire entanglements. More than 350,000 shells were expended before the first soldiers went over the top. However, some British units were able to advance only 914m (1000yds) before being cut to shreds by machine gun fire.

One German in the trenches at the Somme remembered:

'At 7.30 a.m. the hurricane of shells ceased as suddenly as it had begun. Our men at once clambered up the steep shafts leading from the dugouts to daylight and ran singly or in groups to the nearest shell craters. The machine guns were pulled out of the dugouts and hurriedly placed in position, their crews dragging the heavy

The crew of a heavy coastal weapon behind its emplacement. Guns that were inadequate for use in the field were relegated to coastal defence roles.

The United States Army were largely equipped with French and British artillery when they first entered World War I because they were so poorly prepared.

ammunition boxes up the steps and out to the guns. A rough line was thus rapidly established. As soon as the men were in position, a series of extended lines of infantry were seen moving forward from the British trenches. The first line appeared to continue without end from right to left. It was quickly followed by a second line, then a third and fourth. They came on at a steady pace as if expecting to find nothing alive in our trenches.'

Although artillery possessed tremendous destructive power, the Somme demonstrated its limits. The German troops were deeply entrenched in bunkers, some 9m (30ft) deep and reinforced by natural limestone formations. The Somme rapidly degenerated into slaughter. During more than four months of fighting, more than 1,265,000 men were killed or wounded.

An unprepared America

When the United States entered World War I in April 1917, its army was woefully unprepared to provide its own modern artillery. Therefore, with the arrival of the American Expeditionary Force in Europe, US units were largely equipped with French and British artillery. In addition, the US military establishment sought licensing agreements with foreign manufacturers to produce weapons for its army. The most prevalent arms in use by American artillery units included the ubiquitous French 75mm gun, the British 18-pounder and heavier weapons such as the French 155mm GPF gun and the 155mm Schneider and 240mm

howitzers. British heavy weapons such as the 8in and 9.2in howitzers were also in common use.

American forces also adopted a light 37mm French cannon, which was redesignated the 37mm Model 1916, firing a shell of just over 0.54kg (1.19lb) a distance of slightly more than 2377m (2600yds). Toward the end of the war, artillery of American manufacture had begun to arrive on the battlefields of France, but the numbers were relatively insignificant. By November 1918, more than 1800 French-built 75mm guns had been deployed with American forces, while only 143 75mm guns of American manufacture, rechambered to accept the French ammunition available, were in service.

American reliance on the artillery of its allies is also illustrated by the fact that the 120mm Model 1906 gun was the only field piece designed and manufactured in the United States to reach the battlefields of Europe in anything approaching significant numbers. Weighing about 1.8 tonnes (2 tons), the 120mm gun fired a 27kg (60lb) shell at a maximum range of 11,000m (12,140yds). Its use was limited because of ammunition shortages, and the weapon was notable for its unusual carriage, which rolled forwards on wooden wheels with rubber tyres shrunk to fit.

In *Doctrine Under Trial: American Artillery Employment in World War I*, Mark E. Grotelueschen describes the state of the US Army's artillery arm as 'professionally dormant, unprepared … obsolete… In short, when America joined the war, the whole of its army, and especially its field artillery branch, was too small, devoid of any applicable combat experience, insufficiently trained, and in possession of a doctrine that did not appear in any way suited to the daunting military operations that lay in its future in Europe.'

Indeed, American commanders, for the most part, declined to embrace the lessons that their allies had learned at all too high a price during the preceding three years of war. The Doughboys were to learn the costly lesson for themselves at Belleau Wood, where their initial attacks were made without the support of artillery – the casualties were predictably horrendous. Another incident of such an ill-advised assault occurred at Soissons. Slowly, revised tactics were adopted, and the American artillery units contributed to successful operations at St Mihiel, Vaux and Mont Blanc.

The American experience in World War I resulted in a post-war struggle for supremacy among those who remained convinced that infantry employed in open combat would win the day and those who asserted that the harsh reality of the battlefields of the future would require a potent and efficient artillery corps.

Still, units such as the 119th Field Artillery Regiment, a federalized Michigan National Guard unit, acquitted themselves well. Their commander, Colonel Chester B. McCormick, praised the efforts of the regiment:

'The latter part of July [1918], you were rushed into the Second Battle of the Marne referred to as the Marne-Aisne Offensive. You were suddenly confronted with one of the most severe tests of your

ORDNANCE QF 18PDR GUN

Calibre: 83.8mm (3.3in)
Weight in action: 1284kg (2831lb)
Gun Length: 29.4 calibre: 2.46m (96.85in)
Elevation: -5° to +16°

Traverse: 8°
Shell type & weight: shrapnel; 8.4kg (18.5lb)
Muzzle velocity: 492m/sec (1614ft/sec
Maximum range: 8700m (9515yds)

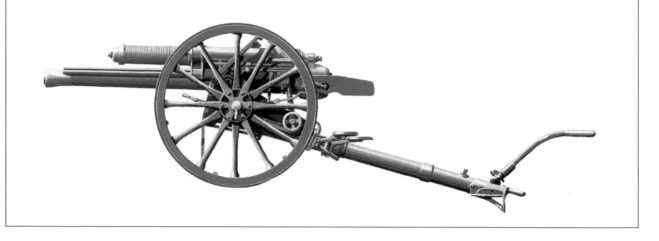

career. With new animals and inexperienced drivers, you were forced to march for five days to the vicinity of Chateau Thierry. On account of the shortage of artillery harness, the regiment was compelled to drag 16 American caissons loaded with ammunition this entire distance. To save the animals, everyone except drivers were compelled to walk and carry full pack for which you had no previous training.'

McCormick also described the regiment's contribution to the capture of Chateau de Auble and Fismettes later that summer. 'Here the gallant and courageous conduct of your gun crews, which time and again were totally destroyed by enemy shell fire, demonstrated that the rigid discipline and details of your early training had not

been without avail; the test came and you met it without faltering.'

In an interesting aside, Battery E of the 11th Field Artillery Regiment staked a claim to firing the last shot of World War I at 11 a.m. on 11 November 1918. The unit's favourite 155mm cannon, nicknamed Calamity Jane, was loaded with a 43kg (95lb) shell, and fired as the second hand of the battery commander's watch swept past 12.

The coming of the war

In 1914, the army of Kaiser Wilhelm II was the most heavily gunned, in terms of artillery, of any European combatant. Pulled inexorably into the vortex of world war through a combination of

American artillery and supply wagons struggle to negotiate a dirt road that has been turned into a quagmire by rains in the spring of 1918.

treaty obligations and its own sabre-rattling, the German nation was already noted for its military prowess and its imperialism. A modern, unified Germany had come into being just decades earlier, after centuries as fragmented fiefdoms. It was forged in large part by the will of Otto von Bismarck, the Iron Chancellor who had made no secret of his country's ambitions to achieve its own 'place in the sun'. Thus the clash of arms that began in the Balkans had seemed, to many, inevitable.

German military doctrine early in the war was dictated by the plan advocated by Field Marshal Count Alfred von Schlieffen, chief of the German General Staff from 1891 to 1905. Schlieffen envisioned a rapid German offensive through lower Holland and across neutral Belgium into northern France. When hostilities broke out, little consideration was given to the prospect of determined Belgian resistance, and the defenders of the fortress city of Liége showed up the naïve optimism of the Germans. The invaders had made no provision to transport heavy siege artillery during their advance.

Liége guarded the crossings of the River Meuse, and the major bridge spanning its 274m (300yds) distance had been partially destroyed. When the Germans reached the Meuse in early August, they repeatedly attempted to bridge the river but found their efforts rebuffed by accurate fire from Belgian batteries. For three days, the Germans were held at the Meuse, thwarting their timetable. The city itself was defended by 12 imposing forts, the largest being triangular in shape with

both heavy and quick-firing artillery pieces, some of which were mounted in revolving steel turrets. Many of these were also mounted on disappearing fixtures, which could rise up to firing position, discharge their round, and then recoil behind the fortress walls, out of sight of observers and protected from counter-battery fire. Crewmen would then place the weapon on target for the next round, usually through reflecting or telescopic sights.

The Germans were taken aback by the ferocity of the Belgians, although the eventual defeat of the defenders appeared to be a foregone conclusion since neither Great Britain nor France planned to come to their relief. The invaders thus delivered an ultimatum for the surrender of the city – which was rejected out of hand by the Belgians.

When German infantry moved forward on 7 August, they sustained serious casualties from Belgian guns and a profusion of land mines. Nevertheless, the assaults continued. Gerald Fortescue, a correspondent for the London *Daily Telegraph*, described a nocturnal assault by German guns:

'Half-an-hour before midnight, a furious bombardment against the southeast forts opened. High explosive shells burst with brilliant flashes and sharp uproar on the very glacis of the forts; a

HEAD TO HEAD: *Canon de 105mm Schneider Mle 1913*

The French 105mm Schneider cannon was manufactured as a result of an initial collaboration between French and Russian designers. Originally a 107mm weapon, it was subsequently modified to accept the French 105mm shell. As the war progressed, the gun was ordered in large numbers by the French government.

Canon de 105mm Schneider Mle 1913

Calibre: 105mm (4.13in)
Weight in action: 2300kg (5070lb)
Gun length: 28.4 calibre: 2.98m (117.3in)
Elevation: 0° to +37°
Traverse: 6°
Shell type & weight: HE; 16.0kg (35.27lb)
Muzzle velocity: 550m/sec (1805ft/sec)
Maximum range: 12,700m (13,890yds)

STRENGTHS

- Effective against fortifications
- Interrupted screw breech
- Heavy shell

WEAKNESSES

- Limited deployment early
- Restricted traverse
- Difficult transport

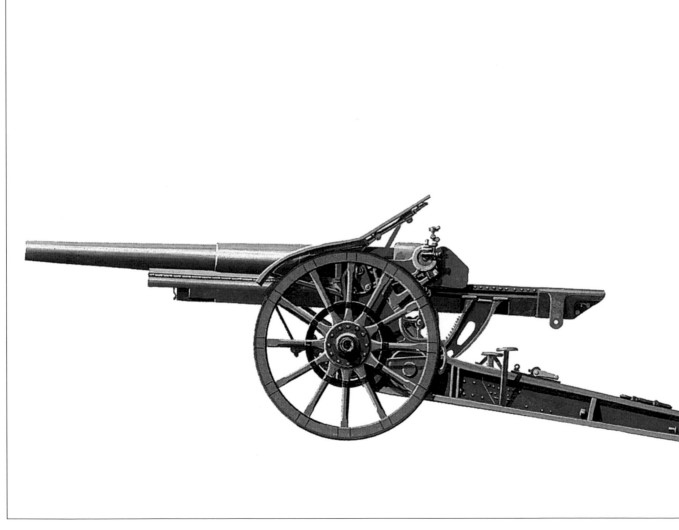

VERSUS   *15cm schwere Infantriegeschütz (sIG) 33*

Development of precursor of the 15cm infantry cannon began in the wake of World War I. The German Army requirement was for a heavy howitzer to complement its light field weapons. The howitzer could be towed by vehicle or horse, and was later mounted on tracked carriages.

15cm schwere Infantriegeschütz (sIG) 33

Calibre: 150mm (5.9in)
Weight in action: 1750kg (3,858lb)
Gun length: barrel 1.65m (5.41ft)
Elevation: 0° + 73°
Traverse: 11.5°
Shell type & weight: HE; 38kg (83.8lb)
Muzzle velocity: 240m/sec (787ft/sec)
Maximum range: 4700m (5140 yards)

STRENGTHS

- Versatile mounting
- Heavy field calibre
- Fired a variety of ordnance

WEAKNESSES

- Extreme weight
- Limited anti-tank effectiveness
- Protracted development

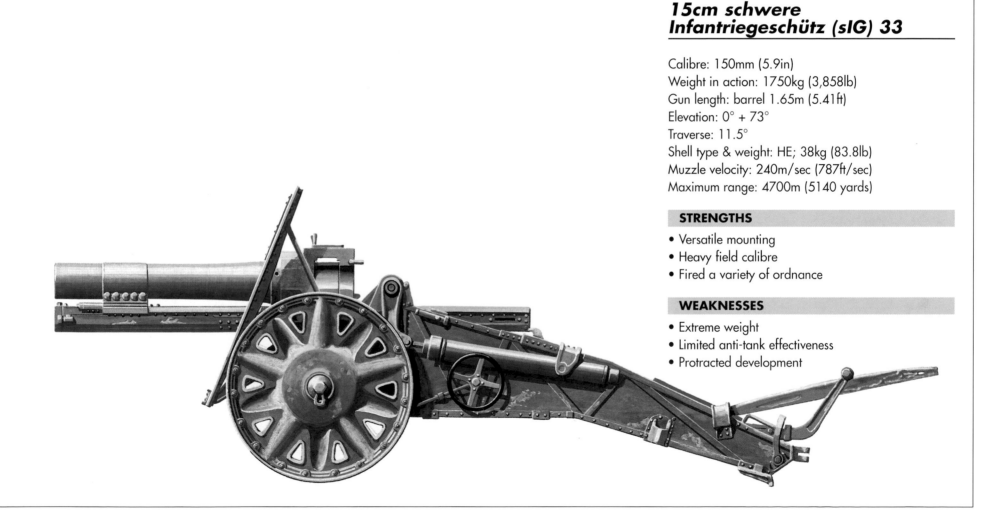

storm of shrapnel broke upon the trenches. The forts replied with energy. The city shook under the thunder of the combat.'

Not until 8 August did the first two forts, Barchon and Evegnée, fall to the Germans, under the command of General Erich von Ludendorff, whose role in Germany's conduct of the war was destined to become more prominent. However, it was apparent that the most strongly defended fortifications were impervious to the light German artillery. Therefore, the Germans reluctantly brought forward massive 280mm, 305mm and even 420mm howitzers, which had been intended for use during the anticipated siege of Paris. When these behemoths arrived, the issue was no longer in doubt. The first of the 420mm howitzers arrived on 12 August, and by 12.30 p.m. the following day, Fort Pontisse was in ruins. During the next two days, six more forts were pounded into submission. The last of these was Fort Loncin, which was wracked by a devastating explosion when a shell detonated its ammunition magazine.

The Belgian commander, General Gerard Leman, was inside Fort Loncin when the fatal shell struck home. 'That I did not lose my life,' he wrote to the exiled king of Belgium while in German captivity, 'is due to my escort, who drew me from a stronghold while I was being suffocated with gas from exploding powder. I was carried to a trench, where I fell.'

Big Bertha

The 420mm siege weapons brought forward at Liége were produced by the famed German arms manufacturer Krupp, located in Essen. Nicknamed Big Bertha, after the wife of the firm's chief, Gustav Krupp, these howitzers were officially designated

Kaiser Wilhelm II's desire that Germany should have its 'place in the sun' in terms of creating an empire contributed to the coming of World War I.

the Model L/14 and fired a shell weighing 820kg (1808lb) a distance of up to 12,000m (13,123yds), with a maximum elevation of 80°∞and an approximate rate of fire of 10 rounds per hour. Six of these howitzers were available at the beginning of the war, and they were transported either by railway or on a wheeled carriage. Aside from being a weapon of great destructive power, the Big Bertha was also a tool of German propaganda, and it gained a fearsome reputation among Allied troops. One Big Bertha actually survived the war and was moved from the Krupp proving ground to Sevastopol in the Crimea for use during World War II.

In April 1915, an assemblage of 92 heavy German howitzers bombarded the Belgian town of Ypres, and the 420mm Big Bertha fired its shells in pairs from the shelter of the nearby Houthulst Forest. The official British history of World War I noted that the shells 'travelled through the air with a noise like a runaway tramcar on badly laid rails.'

The development of the 420mm howitzer had begun in 1900 during Krupp's experimentation with a 350mm weapon. Eight years later, the German Army requested a larger version, and the response in 1912 was a 159-tonne (175-ton) howitzer built in five sections for transportation by rail to its assembly point. Later, a more practical 39-tonne (43-ton) howitzer was introduced, which could be moved in the field. It was this weapon that gained the Big Bertha moniker. The mobile Big Bertha was transported in sections by Daimler-Benz tractors, and a 200-man crew required six hours of strenuous labour to ready it for a firing mission.

The Krupp dynasty

The history of the Krupp family and its arms manufacturing parallels that of an imperial and nationalistic Germany. Friedrich Krupp established a steel foundry in the city of Essen in 1811, and his son, Alfred, was making

Sporting their jaunty headgear, Belgian soldiers march towards the front lines to battle the German invaders. Heavy fighting occurred in Belgium during World War I.

cannon for several armies by 1850. Alfred came to be known as the 'Cannon King' and as 'Alfred the Great', and half of the company's production was devoted to arms manufacturing by 1890. With the dawn of the twentieth century, Krupp was the largest industrial concern in the world.

When Alfred died in 1887, his son Friedrich Alfred gained control of the company and maintained a close relationship with Kaiser Wilhelm II. In addition to artillery, he built the Maxim machine gun and the Diesel engine in his facilities, and was also involved in the beginning of the German Navy's U-boat programme. The Krupp labour force more than doubled during Friedrich Alfred's tenure to 43,000.

Friedrich Alfred died in 1902, reportedly of a stroke but possibly by his own hand, and his daughter, Bertha, inherited the Krupp family fortune. The Kaiser chose her husband, Gustav Krupp von Bohlen und Halbach, who began using the surname of Krupp after the marriage in order to perpetuate the family name.

BIG BERTHA

Calibre: 420mm (16.53in)
Weight in action: 43,285kg (16.53in)
Gun length: 14 calibre: 5.88m (19.3ft)
Elevation: +40° to +75°

Traverse: 4°
Shell type & weight: HE; 820kg (1807lb)
Muzzle velocity: 425m/sec (1394ft/sec)
Maximum range: 9375m (10,252yds)

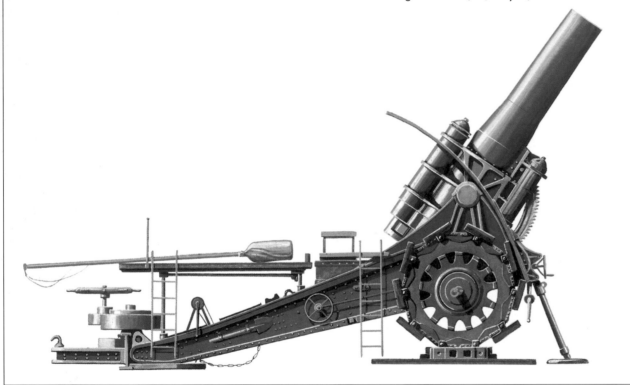

'When World War I broke out, Krupp was the biggest industrial firm on the Continent, with 82,500 workers…' stated a *Time* magazine story on the resurgent company, published in 1957.

'After the German armies were driven back, the victorious Allies chopped off half of Krupp's steel-making capacity, carried the equipment away, destroyed 2,000,000 machines and tools. But they could not destroy the spirit of Krupp's workers, who halted the dismantling process by going on strike during the French occupation…'

The demand for arms had created rapid growth, but Allied concerns following the signing of the Treaty of Versailles redirected the company into the production of heavy machinery and the development of new metal alloys.

The Nazi era facilitated a return to the production of armaments. Although Gustav initially opposed Hitler, he eventually came to be one of the Führer's many admirers. Gustav's son, Alfried Krupp von Bohlen und Halbach, participated with his father in the active rearming of the German war machine during the 1930s. The affairs of Krupp and the affairs of state became thoroughly intertwined, and more than 100,000 slave labourers from occupied countries toiled in Krupp factories. Hitler once thundered to a huge gathering that the youth of Germany must be 'as hard as Krupp steel'.

At the end of World War II, the ageing Gustav was clearly demented; however, both he and Alfried were arrested. Alfried stood trial and was convicted of the abuse of forced labour. Sentenced to 12 years in prison and stripped of his ownership of the firm, Alfried was released when his conviction was overturned in a controversial decision in 1951. More than 70 per cent of the Krupp manufacturing capacity had been destroyed or dismantled by the Allies during World War II, but in 1953 Alfried was allowed to rejoin the company.

In post-war Germany, Krupp refocused on heavy machinery and engaged in railroad production. The rejuvenated company produced Germany's first functioning nuclear plant in 1967. Alfried died that same year, and in 1968 the company became a corporation no longer controlled by the family. One of the greatest industrial dynasties in history had finally come to an end. Merged with Thyssen AG in 1999, the company is today known as

Right: The famed Krupp armaments manufacturing company produced a tremendous amount of ordnance for the German Army during World War I.

HEAD TO HEAD: *Ordnance QF 4.5in Howitzer* VERSUS

Introduced in 1909, the Ordnance QF 4.5in howitzer incorporated valuable lessons learned during the Boer War. An improved version went into production in 1917. The QF designation stands for quick firing, and more than 3000 of the weapons were produced during World War I.

Ordnance QF 4.5in Howitzer

Calibre: 114mm (4.5in)
Weight in action: 1370kg (3020lb)
Gun length: 14.3 calibre: 1.63m (64.17in)
Elevation: -5° to +45°
Traverse: 6°
Shell type & weight: HE; 15.90kg (35.0lb)
Muzzle velocity: 313m/sec (1026ft/sec)
Maximum range: 6400m (7000yds)

STRENGTHS

- Rapid rate of fire
- Excellent against trenches and in barrage fire
- Slide block breech

WEAKNESSES

- Limited effectiveness against heavy fortifications
- Somewhat limited range
- Minimal availability early in World War I

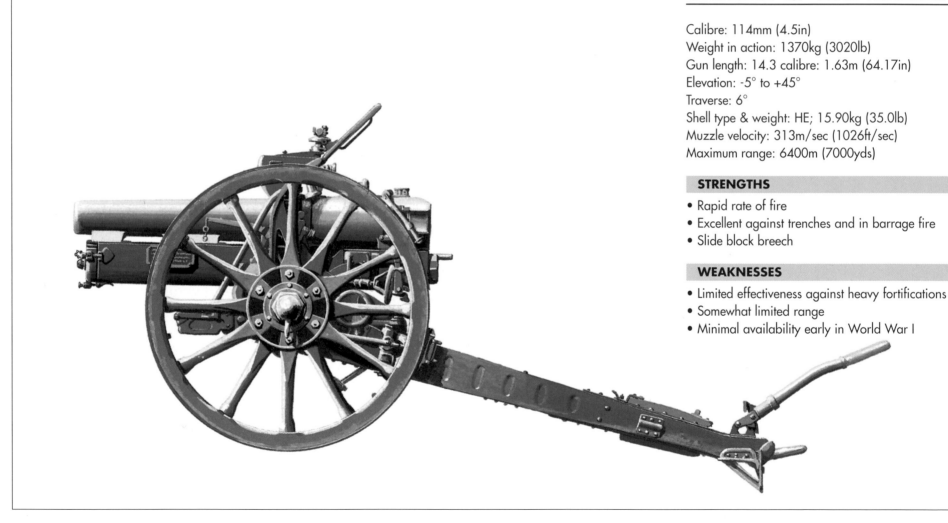

149mm Skoda Model 14 Howitzer

Of Czech design and produced primarily for the armed forces of the Austro-Hungarian Empire, the 149mm Skoda Model 14 Howitzer was also utilized a generation later by Hungarian troops on the Eastern Front during World War II.

149mm Skoda Model 14 Howitzer

Calibre: 149mm (5.86in)
Weight in action: 2765kg (6095lb)
Gun length: 14 calibre: 2.08m (81.88in)
Elevation: -5° to +70°
Traverse: 8°
Shell type & weight: HE; 42kg (92.6lb)
Muzzle velocity: 350m/sec (1148ft/sec)
Maximum range: 8000m (8750yds)

STRENGTHS

- Well designed
- Good elevation
- Solid mobility

WEAKNESSES

- Limited deployment
- Relatively short range
- Limited crew protection

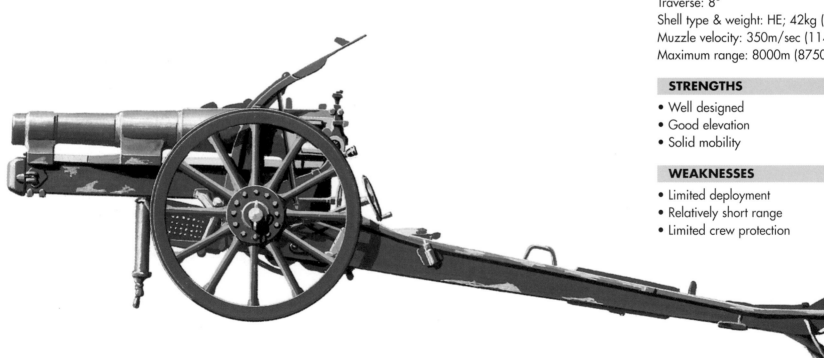

Alfred Krupp, the patriarch of the German armaments manufacturing company, became known as the Cannon King and Alfred The Great in the late nineteenth century.

ThyssenKrupp AG, an industrial conglomerate with interests in numerous fields.

The Whizz Bang

During the course of the war, Allied infantrymen came to refer to the report of any German artillery shell as a 'whizz bang', but the term was originally applied to the sound of the 77mm field gun, which was the most common artillery weapon in the German arsenal in 1914. More than 5000 of these were with German Army units in the beginning, and the 77mm gun was used throughout the conflict. The gun was capable of firing a 6.8kg (15lb) shell a maximum distance of 7800m (8530yds).

The original 77mm design had been introduced in 1896, but the following year saw the debut of the French 75, which rendered the German cannon obsolete. A major redesign effort followed, and the firms of Krupp and Rheinmetall produced the 77mm Feldkanone 96 n.A., the initials identifying it as the 'new model'. The improved version featured a sliding wedge breech and incorporated a recoil system, which its predecessor had completely lacked. Considering German military doctrine of the day, which meant swift manoeuvre and the deployment of both infantry and artillery, the 77mm gun performed its role well. While static trench warfare robbed the weapon of its advantage of mobility and Allied guns had higher rates of fire and delivered heavier shells, the 77mm often compensated for its shortcomings with sheer weight of numbers. On the Eastern Front, it proved valuable in keeping up with rapidly advancing infantry during offensive operations.

The challenges with the 77mm Feldkanone 96 n.A. were readily apparent, and by 1916 Rheinmetall had introduced the 77mm Feldkanone 16, which incorporated a longer barrel with improved rifling and a higher breech volume. The new weapon was popular with the troops in the field, who were now better able to compete with the Allied weapons as the gun's range improved to 9100m (9952yds). Rushed into production and then into the line, the 77mm Feldkanone 16 was initially plagued by a number of operational problems, including the fragility of some of the weapon's smaller parts and a decreased rate of fire. One deadly problem, which involved the bursting of shells in the barrel of the gun, was eventually traced to the type of ammunition being used and the fact that shortages of such vital materials as copper forced the manufacturer to resort to less reliable alternatives.

The primary long-range cannon of the German Army in 1914 was the Krupp M04 100mm gun. Given its primary tasks of suppressing enemy artillery fire and bombarding distant targets, military commanders requested an improved version during the first year of fighting. By 1915, the M14 was in service, and more than 700 of them would eventually be delivered. The M14 improved

Misleading Names

While the weapons are described based upon the weight of the projectile fired, the names are somewhat misleading. The round shot fired by the 4-pounder actually weighed about 4.3kg (9.5lb), and the round shot fired by the 6-pounder was about 6.9kg (15.2lb). Very similar in range, the 4-pounder maximum was approximately 3450m (3773yds), while the 6-pounder could reach targets roughly 3440m (3762yds) distant. Both weapons used bagged charges to propel their rounds.

SKODA 305MM HOWITZER

Calibre: 305mm (12.0in)
Weight in action: 20,000kg (19.68 tons)
Gun length: 14 calibre: 4.26m (13.97ft)
Elevation: +40° to +70°

Traverse: 120°
Shell type & weight: HE; 380kg (838lb)
Muzzle velocity: 340m/sec (1115ft/sec)
Maximum range: 12,000m (13,125yds)

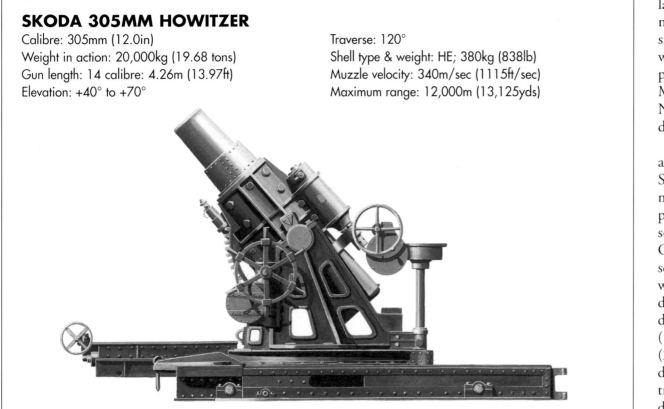

the range of its predecessor from 10,200m (11,155yds) to 13,100m (14,326yds), commonly firing a 6.4kg (14lb) shell. Weighing slightly more than 2.7 tonnes (3 tons), the M14 could be moved by horse-drawn transport systems and deployed as a single piece. In the autumn of 1917, the first of 192 weapons with further improvements reached the battlefield. Designated the M17, the lengthened barrel of this version increased the weapon's range to 16,500m (18,044yds). The M17, however, was somewhat heavier than the M14, requiring the barrel to be removed and transported

separately when relocated by horses. When the war ended, Germany was not allowed under the Treaty of Versailles to retain any of its M17s. However, a number of them were secretly saved from the scrapyard and in use with the German Army as late as 1939.

The Krupp M1913 88mm SK L/45 is worthy of note because it was the first German artillery pieced sized to 88mm, making it, in essence, the progenitor of the 88mm gun of World War II renown. Originally designed as a torpedo boat weapon for the German Navy, the gun was found

lacking in firepower for shipboard service and was more often used in coastal defence. Its vertical sliding breech block distinguished it as a naval weapon, and often the guns were crewed by naval personnel. Early in the war, a number of the M1913 88s were deployed along the coast of the North Sea. They were particularly effective as a deterrent to Allied aircraft.

Originally manufactured by Krupp as a naval and coastal defence gun, the 170mm Schnelladekanone (fast-loading cannon) – nicknamed Samuel when rail-mounted – was placed in service on carriages that had been somewhat hurriedly designed. The heaviest German gun of naval origin to be pressed into service on a movable carriage, the 170mm cannon weighed just over 2268kg (5000lb) and was broken down into three pieces for transport – which was difficult even then. It was capable of firing a 62.8kg (138.5lb) shell a maximum distance of 24,000m (26,247yds). However, its usefulness was limited due to extensive bedding, which affected its traverse. About 60 of these large weapons were used during the war, a number of them converted to the rail-mounted variant.

The 105mm field howitzer 98/09 was widely used by the German Army throughout the war, and in 1914 more than 1200 of them were in service. Like the 77mm gun, it had originally been designed without an internal recoil system, and its rigid carriage made it obsolete by the turn of the century. In 1909, the redesigned weapon with a single motion, wedge breechblock was accepted by the army. Its maximum range of 6300m (6890yds) and relatively high rate of fire made it popular among the troops in the field. With the advent of trench warfare, it became a mainstay of the German

The firepower of a single 105mm howitzer is evident in the muzzle blast of this cannon firing from a concealed position along a wooded front.

though it was not a very accurate weapon its firepower resulted in an extensive zone of destruction. The 250mm mortar weighed 628kg (1385lb) and could be fired across 2600m (2843yds) at a rate of 20 per hour.

The horror of Verdun

'Germany can expect no mercy from this enemy, so long as he retains the slightest hope of achieving his object,' remarked General Erich von Falkenhayn, chief of the German General Staff, in response to the resolve of Great Britain to prosecute the war in late 1915. Therefore, Falkenhayn reasoned that the best hope for a German victory was to deal a deathblow to France. He continued:

'The strain on France has almost reached breaking point – though it is certainly borne with the most remarkable devotion. If we succeeded in opening the eyes of her people to the fact that in a military sense they have nothing more to hope for, that breaking point would be reached and England's best sword would be knocked out of her hand … we can probably do enough for our purposes with limited resources. Within our reach behind the French sector of the Western Front there are objectives for the retention of which the French General Staff would be compelled to throw in every man they have. If they do the forces of France will bleed to death – as there can be no question of a voluntary withdrawal – whether we reach our goal or not.'

Verdun was to be the focal point of a German offensive to knock France out of the war. However, the capture of the city would be no easy task. The reduction of 20 or more heavily fortified strong points, and at least another 40 supporting fortifications, was a daunting prospect. Although

artillery on the Western Front with effective plunging fire. Although two newer 105mm howitzer models were introduced during the war, neither eclipsed the Model 98/09 in numbers or widespread use.

During the course of the war, the German Army deployed heavy calibre mortars, called *minenwerfer* (mine throwers). These cumbersome weapons were valuable due to their increased explosive payload, which was allowed due to the thinner shell casing. For example, the 250mm version of the weapon delivered nearly 47kg (104lb) of high explosive,

only slightly less than 250 rounds of 77mm ammunition. Therefore, the weapon could be tremendously destructive when employed against bunkers or trenches. At the beginning of the war, 44 of these muzzle-loading mortars were available, and they were used during the reduction of the fortresses at Liége. A complement of 21 men was required to move the mortar into position, and

some of the guns at Verdun had been removed earlier in the war and sent to other areas where they were desperately needed, the French were certain to mount a stiff defence.

The most imposing of the major forts, Douamont, housed a 155mm cannon in a retractable turret, which enclosed the entire weapon and did not allow the barrel to noticeably protrude. It was counterweighted so that two soldiers could lower it out of sight. A pair of 75mm guns was enclosed in a retractable turret at the centre of the fort, while three well-protected machine-gun cupolas were situated in steel domes. To the rear, the five-sided structure was protected by three casemated 75mm guns. Along the outer edge of a protective ditch, 37mm cannon and machine guns were positioned in pillboxes. To observers, the position appeared virtually impregnable.

On 21 February 1916, the Germans began their attack with a nine-hour bombardment by more than 500 guns, including the 420mm Big Berthas and 380mm and 305mm weapons. More than 80,000 shells were expended. Within three days, the Germans had taken the outer defences of the city, and the fortifications had been hammered but were holding out. The following day, the unthinkable happened. Fort Douamont fell to the Germans. The French commander, Field Marshal Joseph Joffre, was replaced by General Philippe Pétain, who made his famous pledge 'They shall not pass.'

French soldiers rest during a brief respite towards the battlefield at Verdun. The battle came about as the Germans attempted a decisive offensive in the West.

The situation became critical for the French, but re-supply efforts along a single road from Bar-le-Duc were successful, and the route came to be known as the Sacred Way. As the battle wore on in March, Fort Vaux, not far from Douamont, was lost and retaken 13 times. British correspondent Lord Northcliffe visited the battlefield at Verdun in the midst of the fighting. On 4 March 1916, he wrote:

'The present attack on the French at Verdun is by far the most violent incident of the whole Western War. As I write it is late. Yet the bombardment is continuing, and the massed guns of the Germans are of greater calibre than have ever been used in such numbers… Whatever may be the result of the attack on the Verdun sector, every such effort will result in adding many more thousands of corpses to those now lying in the valley of the Meuse.'

By August, the desperate battle of attrition was beginning to ebb as the Germans were forced to

send reinforcements to the east when the Austrian Front collapsed and the Russians captured 400,000 prisoners. In the autumn, the French launched a counterattack that recaptured forts Douamont and Vaux. By December, the decisive battle was over, and more than 1,250,000 were dead or wounded. The fighting at Verdun between 1915 and 1918 lasted a total of 18 months. Estimates of the number of artillery shells fired run as high as 12 million, or 22,000 per day.

A young French soldier, who had endured the terror of an artillery bombardment at Verdun, recalled:

'When you hear the whistling in the distance your entire body preventively crunches together to prepare for the enormous explosions. Every new explosion is a new attack, a new fatigue, a new affliction. Even nerves of the hardest of steel are not capable of dealing with this kind of pressure. The moment comes when the blood rushes to your head, the fever burns inside your body and the nerves, numbed with tiredness, are not capable of reacting to anything anymore. It is as if you are tied to a pole and threatened by a man with a hammer. First the hammer is swung backwards in order to hit hard, then it is swung forwards, only missing your skull by an inch, into the splintering pole. In the end you just surrender. Even the strength to guard yourself from splinters now fails you. There is even hardly enough strength left to pray to God.'

War in the east

Austria-Hungary, Germany's principal partner among the Central Powers, felt the sting of Serbian artillery early in the war when the Austrian 21st Division was severely mauled at Cer Mountain near the Drina River in August 1914. The Austrians

Verdun's Dead

The Verdun Memorial stands as mute testimony to the ferocity of the fighting. When the struggle was finished, well over a million men had perished in the conflagration. The bones of the dead continue to be unearthed to this day, and many of them are deposited in huge ossuaries on the battlefield.

themselves possessed capable, modern artillery, much of which was produced at the Skoda Works. Numerous quality artillery pieces were produced by Skoda and sold to the armed forces of other nations as well.

Among the most famous Skoda artillery of the Great War was the 305mm Model 1911 siege mortar, which was deployed by the Germans at Liége. The heavy weapon was capable of firing a 384kg (847lb) shell up to 9601m (10,500yds) or a lighter 287kg (633lb) shell 11,300m (12,358yds).

When the smaller shell detonated, its killing zone extended more than 393m (430yds), and its impact crater was nearly 8.2m (9yds) deep and wide. Serviced by a crew of 12, it could fire 10 rounds in an hour. The weapon was transported in three sections by a Skoda-Daimler tractor and could be set up to fire in less than an hour. In 1916, an upgraded version extended its range to more than 12,162m (13,300yds), and by the end of the war 40 of these mammoth weapons were in service. Skoda also produced 75mm, 100mm and 150mm artillery pieces, which saw service in numerous armies during the war.

Following their resounding defeat in the Russo-Japanese War, the army of Czarist Russia was obliged to modernize, and during the decade from 1904 to the beginning of World War I, artillery was purchased from both France and Germany. From the French, the 120mm Schneider howitzer was rebored to accept standard Russian 122mm ammunition and renamed the M10. Another 122mm howitzer, the M09 was of Krupp manufacture. The 91mm M16 trench mortar, allocated to the Russian Army in 1916, was manufactured in Helsinki, Finland, by the firm of AB Maskin & Bro.

One Austrian soldier remembered the power of the Russian mortars:

'At last, on the fourth day, they brought their heavy mortars against us. They were splendidly aimed, but the holy saints guarded us. The shells all burst within a space of about 100 square metres where my guns were standing, killing five men and four horses. Two shells fell exactly on the guns, but did not explode. A gun shield was completely pierced, the wheels of the ammunition carts shot to pieces – but after four days we were able to carry off one gun with whole skins. A shell burst in my observation stand. A captain of the General Staff sitting near me was killed, and a first lieutenant of the General Staff was severely wounded. The battle is still going on. The Russians are showing themselves very obstinate, continuing to persist in the face of slaughter.'

Austrian artillerymen demonstrate the servicing of their artillery piece. The Austrians were on the receiving end of deadly Serbian artillery fire at Cer Mountain.

Fritz Kreisler, a violinist in peacetime, remembered serving in the trenches with the Austrian Army and narrowly escaping a Russian artillery barrage:

'We marched on until the command was given for us to deploy, and soon afterwards the first shrapnel whizzed over our heads. It did no harm, nor did the second and third, but the fourth hit three men in the battalion in the rear of us. Our forward movement, however, was not interrupted, and we did not see or hear anything beyond two or three startled cries. The next shell burst right ahead of us, sending a shower of bullets and steel fragments around. A man about 20 yards to the right of my company, but not of my platoon, leaped into the air with an agonizing cry and fell in a heap, mortally wounded. As we were advancing very swiftly, I only saw it as a dream, while running by. Then came in rapid succession four or five terrific explosions right over our heads, and I felt a sudden gust of cold wind strike my cheek as a big shell fragment came howling through the air, ploughing the ground viciously as it struck and sending a spray of sand around.'

The workhorse of the Russian Artillery corps during World War I was the Putilov 76.2mm M02

gun. Thousands were produced by the Putilov Company in St Petersburg. The weapon utilized a screw breech and fired a shell weighing 7.4kg (16.3lb) a maximum range of just under 9601m (10,500yds). It was used during the Russian Civil War and later in World War II.

The use of chemical agents by both sides during World War I is well documented; however, each blamed the other as the first to do so in violation of The Hague Conventions. In 1915, Germany cited French use of turpinite and chloroacetone as justification for its own employment of chlorine gas in cylinders at Ypres in April of that year. Mustard and phosgene gases were also used during World War I, and most artillery pieces were capable of delivering the chemical agents.

As both sides used gas, the technology of delivery progressed, and the British eventually developed the Livens Projector, which consisted of a steel tube set atop a base plate. Fired electrically, batteries of these weapons could send a number of large but thin-skinned gas bombs toward the enemy to create a sizable cloud. The Livens Projector proved to be a better system to deliver gas in quantity than the shells fired by artillery.

Left: A charging British soldier clutches his throat after sustaining a wound from a bullet. Others have donned protective masks as the threat of a gas attack is present.

Contrary to belief, the number of fatalities due to gas attacks during World War I was surprisingly small. Fewer than one in 20 victims of gas actually died, and 93 per cent of gas casualties during the war actually returned to duty. Between 1914 and 1918, the British Army suffered a combined total of 181,000 non-lethal casualties from mustard and chlorine gas, while only slightly more than 6000 died.

The railway guns

On 29 March 1918, a German artillery shell slammed through the roof of the church at St Gervais near Paris, killing 100 people. Tragic though the event was, the source of the bombard-ment was astounding. The round had been fired by the Kaiser Wilhelm Geschutz, also known as the Paris Gun, which had shelled the French capital for the first time a week earlier from the forest area of Coucy, a distance of more than 112km (70 miles).

Manufactured by Krupp, the Paris Gun was the largest artillery weapon constructed up to that time. Serviced by a crew of 80 sailors of the German Navy, the cannon was capable of firing a 95kg (210lb) shell up to 130km (81 miles), with the projectile actually entering the stratosphere during its flight. The gun weighed 232 tonnes (256 tons), and its rifled 210mm (8.3in) barrel was 28m (92ft) long with a 6m (20ft) smoothbore extension.

Wearing gas masks as they man the front line in their trench, soldiers watch for an enemy advance. Poisonous gas was often deployed via artillery shells.

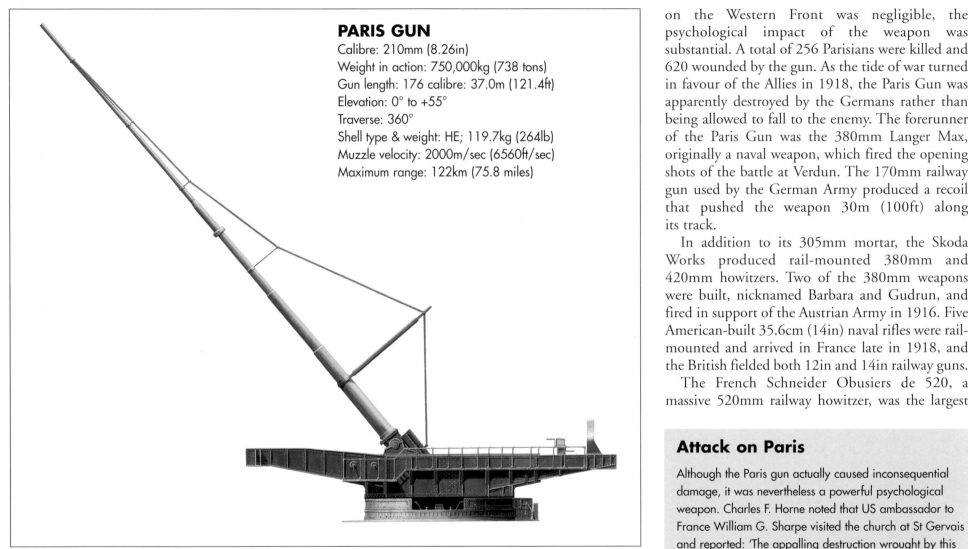

PARIS GUN

Calibre: 210mm (8.26in)
Weight in action: 750,000kg (738 tons)
Gun length: 176 calibre: 37.0m (121.4ft)
Elevation: 0° to +55°
Traverse: 360°
Shell type & weight: HE; 119.7kg (264lb)
Muzzle velocity: 2000m/sec (6560ft/sec)
Maximum range: 122km (75.8 miles)

Due to the high velocity of each shell, the wear on the rifle bore was considerable. After 65 sequentially numbered shells, each progressively larger, were fired, the barrel was rebored to 240mm (9.4in).

From March to August 1918, the Paris Gun reportedly fired 367 shells, about half of which landed within the boundaries of the City of Light. Although the total damage proved relatively insignificant, and its effect on the military situation on the Western Front was negligible, the psychological impact of the weapon was substantial. A total of 256 Parisians were killed and 620 wounded by the gun. As the tide of war turned in favour of the Allies in 1918, the Paris Gun was apparently destroyed by the Germans rather than being allowed to fall to the enemy. The forerunner of the Paris Gun was the 380mm Langer Max, originally a naval weapon, which fired the opening shots of the battle at Verdun. The 170mm railway gun used by the German Army produced a recoil that pushed the weapon 30m (100ft) along its track.

In addition to its 305mm mortar, the Skoda Works produced rail-mounted 380mm and 420mm howitzers. Two of the 380mm weapons were built, nicknamed Barbara and Gudrun, and fired in support of the Austrian Army in 1916. Five American-built 35.6cm (14in) naval rifles were rail-mounted and arrived in France late in 1918, and the British fielded both 12in and 14in railway guns.

The French Schneider Obusiers de 520, a massive 520mm railway howitzer, was the largest

Attack on Paris

Although the Paris gun actually caused inconsequential damage, it was nevertheless a powerful psychological weapon. Charles F. Horne noted that US ambassador to France William G. Sharpe visited the church at St Gervais and reported: 'The appalling destruction wrought by this shell is ... probably not equalled by any single discharge of any hostile gun in the cruelty and horrors of its results... Nearly a hundred mangled corpses lying in the morgues, with almost as many seriously wounded, attested to the measure of the toll exacted...'

calibre artillery weapon of World War I. The howitzer fired a 1406kg (3100lb) shell more than 16km (10 miles) from a railcar of just less than 30.5m (100ft) in length. The entire apparatus weighed 263 tonnes (290 tons).

Lessons to be learned

The emergence of artillery as a dominating presence on the modern battlefield is evidenced by the fact that up to 600,000 artillerymen were serving in uniform at any time in World War I. During the offensive against German entrenchments in Champagne in the autumn of 1918, a total of 1,375,000 shells were fired by the French 75mm field guns alone. Estimates of the casualties inflicted by artillery during World War I are as high as 67 per cent of the total dead and wounded, an astounding increase from the levels in previous wars of approximately 15 per cent.

In *White Heat – The New Warfare, 1914–18*, John Terraine noted, 'The war of 1914–18 was an artillery war; artillery was the battle-winner, artillery was what caused the greatest loss of life, the most dreadful wounds, and the deepest fear.'

The devastation wrought by artillery during World War I can scarcely be described. Certainly, the size, firepower and number of weapons increased as the war progressed, necessitating changes in tactical deployment and setting the stage for an unprecedented degree of specialization in the future. Artillery was fielded in greater numbers than ever before during World War II, which erupted little more than 20 years later.

One of numerous massive railway guns is shown mounted on its huge carriage, which facilitated the dissipation of the recoil when the giant weapon was fired.

THE INTERWAR YEARS

The catastrophic losses, of men and finances, sustained by the nations of Europe during World War I left them in financial straits and in the throes of sweeping political change, and with the spectre of unresolved conflict looming across the continent. The Treaty of Versailles, signed in the stately Hall of Mirrors at the magnificent palace of Louis XIV, settled virtually nothing.

Great Britain was virtually bankrupt, its people lamenting the sacrifice of the 'flower of a generation' in the crucible of total war. France had been bled white and left in a national quandary as political polar opposites assailed its republican form of government. The political landscape in Russia had been dramatically altered with the ascent of the communist Bolsheviks and the murders of the Czar and his family. The Soviet Union was to become a burgeoning political and military force, but for a time it was required to contend with the internal struggle to consolidate power. The empire of Austria-Hungary had been dismembered and the map of Europe redrawn, once again stirring centuries-old grievances and ethnic rivalries. Imperial Germany had ceased to exist, its monarchy a thing of the past and the weak, ineffective Weimar Republic left to contend with soaring inflation and great political unrest.

Left: New field artillery pieces gleam on the factory floor. Lessons learned during World War I were implemented in improved artillery in the 1920s and 1930s.

In the United States, the devastation of world war had been profoundly shocking. Even though the country had been spared the destruction and loss of life on the scale experienced by the Europeans, a growing isolationist sentiment did have a profound impact on the American psyche. At the same time, the rise of militarism in Japan was fostered by the need for precious raw materials to fuel a programme of industrialization, which had begun in the latter part of the nineteenth century. The Japanese bid for pre-eminence in the Pacific placed the nation on a collision course with the United States, which would later erupt in open warfare.

The lessons of the Great War

War had accelerated the pace of development in artillery. For example, it had been quickly determined that shrapnel shells were of little consequence against fixed fortifications or trenches; therefore, most of the warring nations placed emphasis on the high-explosive shells. The changing demands of the battlefield had also given rise to greater specialization of artillery. Coastal defence guns had been a primary focus of artillery designers during the last quarter of the nineteenth century, but many of the guns had been removed from their original mounts during the war and employed in the field. That said, fixed coastal fortifications continued to play a role in the strategic thinking of commanders during World War II.

The aeroplane came of age during World War I, both as a means of finding and fixing targets and as a threat to artillery on the ground, either in an established position or in transit. Therefore, research into effective anti-aircraft guns had been initiated as early as 1909. However, the primary obstacle had been developing a suitable fire-control system that could acquire a moving target hundreds of feet in the air and accurately account for such phenomena as the wind, temperature and even the curvature of the earth. During the 1920s, designers were able to employ a fire predictor, which accounted for certain variables and the inherent characteristics of the gun in order to plot the future position of the aircraft and fire a shell at this 'predicted' position. Work with radar was in its infancy during the interwar years.

The appearance of the tank on the battlefield presented another challenge to the supremacy of artillery, and work on anti-tank weapons began seriously during the 1930s. Early anti-tank guns were light and mobile, principally of no higher firepower than 40mm (1.57in). Although these were effective against the thin skins of early tanks, thicker armour and heavier vehicles necessitated the introduction of more powerful anti-tank guns.

The introduction of time fuses during World War I had given gunners the ability to set the interval between firing a shell and the detonation of the explosive contained within. The result was a shellburst above ground, sometimes just over the heads of exposed enemy troops, raining shrapnel upon these formations and inflicting greater casualties than solid shot or shells equipped with impact fuses. A significant degree of specialization in artillery shells emerged in World War I as well, including smoke shells that could obscure troop movements, gas shells to deliver chemical agents, flare shells for illumination and incendiary shells specifically developed to ignite the highly flammable hydrogen gas that filled German zeppelins.

Mobility had been a constant challenge for field artillery during World War I, and often the ponderous disassembly and limited transportation capability of horse-drawn weapons impeded the progress of an offensive. It was universally recognized that the artillery of the future would have to be mobile, enabling it to keep pace with the tank and other mechanized weapons of war. Therefore, research was begun on self-propelled artillery that could deliver firepower rapidly, creating or exploiting a breakthrough, or halting an enemy movement in its tracks.

For all the innovations it had fostered, the modern battlefield presented new challenges for the artillery as well. As Ian V. Hogg notes:

'The First World War had brought several new lessons before the gunner. Barrage firing with infantry had made greater demands for accuracy than had heretofore been necessary in field gunnery, leading to study of the relationships between the wear of the barrel caused by shooting, and the velocity and accuracy of the shell... The conditions of warfare on the Western Front had posed new problems in communication and control, and had made direct fire at targets seen from the gun a rarity instead of the rule... All these problems, and more beside, were grappled with during the interwar years, though much of the grappling was hampered by the financial stringency which follows any war. It was further hampered by

Aerial photography and observation assumed an important role during World War I, and that role expanded steadily during subsequent years.

the dissolution of many of the design teams and "think factories" which had been forged between 1914 and 1918; for example, much of the pioneer work on the theoretical aspects of anti-aircraft gunnery and fire control had been done by civilian mathematicians and physicists gathered from British universities, together with men of similar calibre who were already in uniform. When the war ended most of the uniformed men were demobilized and the professors returned to their universities to resume their peaceful pursuits – after all, it had been the war to end all wars.'

Struggling with deficiencies

Despite anti-war sentiment and financial constraints, the guns of the Great War had hardly fallen silent when the military establishment of the US Army initiated a formal assessment of its artillery capabilities. The reliance on British and French guns during 1917–18 was a source of embarrassment for many, and in 1919 the Calibre Board, headed by General William Westervelt and popularly known as the Westervelt Board, met for the first time in France. Four months later, the board submitted a formal report, which stated bluntly, 'Every item of hardware of war need[ed] improvement.'

As Hogg relates:

'In 1918, every nation had its artillery staffs at work, drawing up specifications of the next generation of guns, basing its demands on the lessons just learned. The most exhaustive study was probably that mounted by the US Army… This board quizzed British, French and American artillerymen to determine what was good, what was bad, what was lacking in their wartime equipment, and what they would like to see in the future. In 1919 the Board published its findings and made recommendations as to calibre and desirable features of the future artillery of the US Army, which were to be the fundamental guidelines for the subsequent 30 years. Indeed, it would not be too far-fetched to say that Westervelt's policies have even had their effect on NATO (North Atlantic Treaty Organization) doctrine…'

The 105mm howitzer

In reviewing the composition of the armies in the previous war, the Westervelt Board recommended the development of a 105mm howitzer as the backbone of the artillery component of US Army divisions. Some difficulty and expense would be encountered in transitioning from the older 75mm weapons obtained directly from the French or built under licence and steadily upgraded during the interwar period. However, it was not until 1939, a full 20 years after the Westervelt recommendation, that the research and development of the howitzer was completed. Subsequently, the 105mm howitzer became the most important field artillery weapon of

Worst to First

In the months following World War I, the US Army recognized its deficiencies in organic artillery and its dependence on weapons of foreign manufacture. Although the period between the publication of the findings of the Westervelt Board and the implementation of many of its recommendations was lengthy, the United States emerged during the interwar years as the world's principal innovator and proponent of field artillery for use in battle. For three decades these findings influenced the development of artillery with the US armed forces.

World War II for the United States. From April 1941 until June 1945, more than 8500 of the guns were produced.

The 105mm howitzer was capable of firing a 15kg (33lb) shell a distance of 11,430m (12,500yds) at a maximum elevation of 65°. Its rate of fire was up to four rounds per minute, and it incorporated a hydropneumatic recoil system and sliding breechblock. Mounted on the M2A2 carriage, the wheeled weapon was primarily towed into position by the army's 2.3-tonne (2.5-ton) truck. Heavy for its calibre, the weapon weighed a hefty 2259kg (4980lb). As World War II progressed, the basic design of the 105mm howitzer underwent continual revisions, such as the M3, which featured a shortened barrel and was deliverable by aircraft. The success of the design is undisputed, and its ancestors remain in service today.

Blistering barrels at Bloody Ridge

The American landings on the island of Guadalcanal in August 1942, marked their first offensive ground effort against the Japanese in World War II. The American foothold on the island was tenuous in the face of repeated attacks by Japanese forces, and a key position defending the only operational airfield on the island, Henderson Field, was under constant attack in mid-September. Commanding a force of 800 men of the 1st Marine Raider Battalion, Colonel Merritt A. 'Red Mike' Edson made a desperate stand along a commanding ridgeline.

Veteran J. R. Garrett remembered:

'The 11th Marines' 105mm howitzers gave good account of themselves in the battle with the heaviest concentration of artillery fire Guadalcanal had seen so far, dropping well-placed barrages into enemy positions just 200 yards from the dug-in Marines. When it was over the Marines' 105 howitzers had fired 1992 rounds into the enemy's ranks. The 75s alone had unloaded more than a thousand. The Japanese suffered 1200 casualties.'

The Marine position held, and the success at 'Bloody Ridge' was due largely to the performance of the artillery. The official Marine Corps monograph on the Battle of Guadalcanal reported:

'A forward observation post [OP] was maintained on the ridge, and fire from the 105mm howitzers was adjusted form that post. Wire communication was successful between the OP and the fire-direction center, being interrupted only when the OP was displacing rearward from its farthest forward position.'

Another author wrote:

'The way it fell it looked as if the artillery lads were trying to burn out their barrels, so fast and furiously did the shells go over the Raiders. Out of this barrage grew an apocryphal story: a Jap officer is supposed to have asked later, upon his capture, to see the 'automatic artillery' we used that night.'

US field and heavy artillery

During the interwar years, the US Army maintained 75mm field cannon and howitzers in its arsenal, and chief among these was the 75mm pack howitzer M1 developed between 1920 and 1927. Designed primarily for use in mountainous terrain and confined areas where towed or heavier pieces could not be accommodated, the 75mm pack howitzer could be taken apart and transported by animals along narrow roadways. Later in World War II, an airborne version of the weapon was developed and designated the M1A1. Nearly 5000 of the 75mm pack howitzer were produced during World War II, and the weapons were found particularly useful during the rugged Italian campaign. Weighing 653kg (1440lb), the 75mm pack howitzer used a sliding breechblock and fired a 6.6kg (14.6lb) projectile a maximum range of 8925m (9760yds) at a rate of up to eight rounds at 30-second intervals.

The 155mm M1918 was essentially the old French Schneider 155mm howitzer built under licence in the United States, and a few of these actually reached American troops in France before the armistice ended World War I. A debate raged in the United States concerning the development of a more modern 155mm howitzer. By the mid-1930s, a renewed commitment to the weapon was realized and the result was the 155mm M1, which became, after the 105mm, the most widely employed field gun of World War II. Between the spring of 1941

During field exercises, US soldiers wrestle a 105mm M2A1 artillery piece into position. The 105mm M2A1 served as the backbone of US artillery in World War II.

37MM ANTI-TANK GUN M3

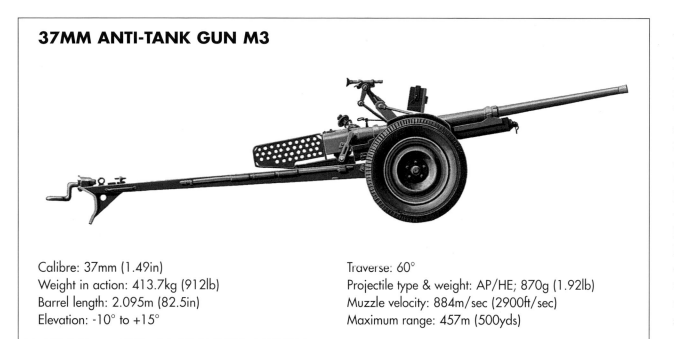

Calibre: 37mm (1.49in)
Weight in action: 413.7kg (912lb)
Barrel length: 2.095m (82.5in)
Elevation: -10° to +15°

Traverse: 60°
Projectile type & weight: AP/HE; 870g (1.92lb)
Muzzle velocity: 884m/sec (2900ft/sec)
Maximum range: 457m (500yds)

and the end of the war, more than 4000 were produced, and the weapon was capable of firing a 43kg (95lb) shell a distance of 14,630m (16,000yds). Utilizing an interrupted thread breech, its rate of fire was two rounds per minute.

Another prominent weapon was the famous 155mm M1 cannon, nicknamed 'Long Tom'. With a distinctive length of 13.7m (45ft), the Long Tom fired a 43kg (95lb) shell up to 24km (15 miles) at a rate of 40 rounds per hour. Its weight of 13,880kg (30,600lb) required a tracked prime mover for relocation in most cases. The weapon utilized an interrupted screw breech, and nearly 1900 were produced between 1940 and 1945.

Primary US heavy artillery pieces with origins during the interwar years were the 8in howitzer M1, 240mm howitzer M1 and the 8in gun. Worthy of

special note is the 240mm howitzer, which weighed more than 27 tonnes (30 tons) and hurled a 163kg (360lb) shell more than 22,860m (25,000yds). Production on this gigantic system began in the autumn of 1942, but its future was very much in doubt.

As Steven J. Zaloga explains: 'The Fifth Army in Italy, the first recipient of the 240mm howitzer in the spring of 1944, was reluctant to accept the new battalions.' The VI Corps commander, Major General John P. Lucas, was 'doubtful of the value of the 240mm howitzer in this country'. In spite of their earlier reluctance, the Fifth Army commanders later dubbed the 240mm howitzer as 'the most generally satisfactory weapon' in service in 1944, and it was popularly nicknamed the 'Black Dragon' by troops.

By 1939, the ageing stock of Great Britain's World War I-vintage 18-pounder field guns and 4.5in howitzers was in need of replacement. British military commanders had been aware of the need for some time, and research on a weapon that could combine the attributes of a field gun and a howitzer had begun during the 1920s. However, financial constraints had prevented a large-scale commitment to such a weapon. The initial Ordnance, Q.F., 25-pounders were essentially heavier guns mounted on the carriages of the old surplus 18-pounders. The carriages had been modernized with pneumatic tyres, and some had been equipped with split trails. A number of these guns, designated Mark I, went to war with the British Expeditionary Force on the continent and were lost during the evacuation at Dunkirk.

Simultaneously, however, the Mark II, engineered on a combined basis including gun and carriage, was being deployed with Commonwealth troops and was destined to become one of the best-known artillery pieces of World War II. The 25-pounder fired a shell of that weight (11.3kg) a range of 12,252m (13,400yds). Its sliding breechblock enabled the gun to fire five rounds per minute.

The fighting in North Africa brought the 25-pounder to prominence. One author related:

'The battles of the Western Desert between 1940 and 1942 against the Axis could be argued to have been the finest hour of the 25-pdr … the arrival of Rommel's Afrika Korps changed the situation almost overnight. Being almost 88mm in calibre itself, the 25-pdr was the only gun capable of challenging the German armour and the battles around and to the east of Tobruk would prove the

HEAD TO HEAD: *8in Howitzer M1* VERSUS

Originally designed during the interwar years in response to the production of heavy artillery in Germany, the 8in howitzer M1 was used during the Vietnam era and by NATO forces in the 1960s. A prime mover or gun tractor was required to relocate the weapon.

8in Howitzer M1

Calibre: 203mm (8in)
Weight in action: 14,380kg (14.15 tons)
Gun length: 25 calibre: 5.08m (16.67ft)
Elevation: -2° to +65°
Traverse: 60°
Shell type & weight: HE; 90.73kg (200lb)
Muzzle velocity: 595m/sec (1952ft/sec)
Maximum range: 16,925m (18,510yds)

STRENGTHS

- Heavy calibre
- Long range
- Excellent accuracy

WEAKNESSES

- Functional obsolescence due to size
- Significant weight
- Lack of mobility

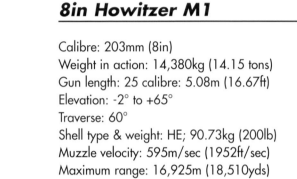

sFH18

Based on a design first conceived during World War I, the 150mm sFH 18 was the primary German heavy howitzer of World War II. Nicknamed the Evergreen, it was the first field cannon capable of firing rocket-assisted ammunition. The weapon remained in service until the 1970s.

sFH18

Calibre: 150mm (5.91in)
Weight in action: 5512kg (12,150lb)
Gun length: 27.5 calibre: 4.125m (13.533ft)
Elevation: -3° to +45°
Traverse: 64°
Shell type & weight: HE; 43.5kg (96lb)
Muzzle velocity: 495m/sec (1624ft/sec)
Maximum range: 13,250m (14,490yds)

STRENGTHS

- Innovative rocket-propelled ordnance
- Self-propelled application
- Pneumatic tyres for ease of transport

WEAKNESSES

- Limited range
- Excessive barrel wear
- Strong recoil

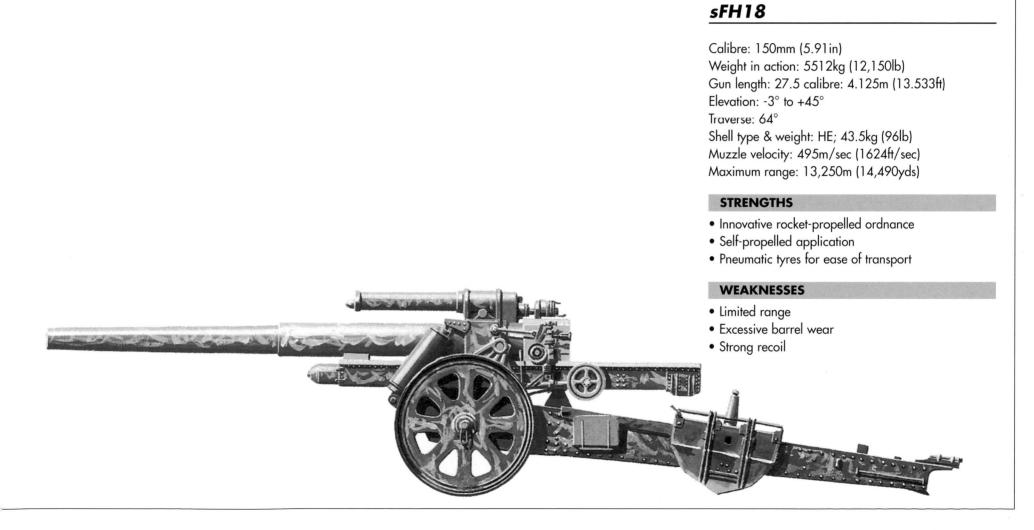

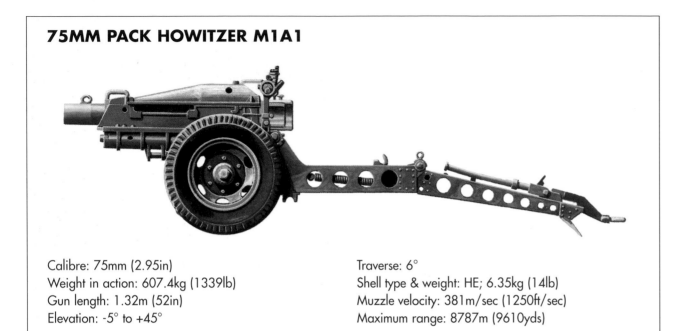

75MM PACK HOWITZER M1A1

Calibre: 75mm (2.95in)
Weight in action: 607.4kg (1339lb)
Gun length: 1.32m (52in)
Elevation: -5° to +45°

Traverse: 6°
Shell type & weight: HE; 6.35kg (14lb)
Muzzle velocity: 381m/sec (1250ft/sec)
Maximum range: 8787m (9610yds)

worth of the gun. The fighting around Tobruk in April 1942 saw the 25-pdr pitted directly against German armour, for example by A/E Battery, 3rd Regiment, Royal Horse Artillery and it proved itself a worthy opponent.'

The weapon gained further acclaim during the decisive battle of El Alamein in October 1942. El Alamein was the turning point in the war in North Africa, and following the Eighth Army victory, the British pursued the German Afrika Korps westwards across thousands of kilometres of desert.

General Bernard Law Montgomery, commander of the Eighth Army, issued his order of the day on 23 October. He told his troops:

'The battle which is now about to begin will be one of the decisive battles of history. It will be the turning point of the war. The eyes of the whole world will be on us, watching anxiously which way the battle will swing. We can give them their answer at once, "It will swing our way." We have first-class equipment; good tanks; good anti-tank guns; plenty of artillery and plenty of ammunition; and we are backed by the finest air-striking force in the world. All that is necessary is that each one of us, every officer and man, should enter the battle with the determination to see it through – to fight and to kill – and finally, to win. If we do this there can only be one result – together we will hit the enemy for "six", right out of Africa.'

To make good on his assertion, Montgomery had amassed more than 800 guns for the El Alamein offensive. During a week of fighting, his artillery fired over a million rounds. The 25-pounder proved efficient and effective. Its employment against

German armour occurred out of necessity following the dismal failure of the weak 2-pounder. Variants of the 25-pounder included an Australian model that could be broken down for transport by pack animals and an airborne or jungle version with a narrow carriage to facilitate its transport in cramped quarters.

Scramble for the heavy artillery

When the British Army needed heavy artillery in 1940, years of inactivity with regard to improvements necessitated that World War I veterans such as the 8in howitzer should be placed into service. One rapidly taken measure was the

British soldiers in the North African desert take cover as German counter-battery fire comes dangerously close to their exposed position.

introduction of the 7.2in howitzer, essentially reboring the old 8in barrels to accept new ammunition. Problems arose with the instability of the carriage when the weapon was discharged, and steel ramps were placed behind the tyres to minimize the roll. Six versions of the 7.2in howitzer were ultimately produced during World War II, and later the American M1 carriage offered additional stability. Weighing 10,387kg (22,900lb), the M I-V versions were capable of firing a 92kg (202lb) shell a range of 15,453m (16,900yds), while the much larger M6, more than 1.8m (6ft) longer and weighing 13,209kg (29,120lb), had an improved range of nearly 18,288m (20,000yds).

Twilight for France

As World War II erupted, the French Army still possessed more than 4500 of its venerable 75mm guns in service with frontline units. When France was overrun in the spring of 1940, many of these weapons fell into the hands of the Germans and were eventually used in the defences of the Atlantic Wall. Interestingly, when the Wehrmacht encountered the Red Army's T-34/76 tank on the Eastern Front, a number of stockpiled 75s were refitted with muzzle brakes and strengthened barrels. They were then pressed into service against the Soviets, designated by the Germans as the 75mm Pak 97/38.

Left: A US-made 155mm cannon, nicknamed Long Tom, thunders against enemy positions in France. It could hurl a shell up to 24km (15 miles) at 40 rounds per hour.

The Schneider 105mm M1913, a cannon of Great War vintage, was sold to a number of European armies between the world wars, including those of Poland and Italy. The Germans admired the weapon, which had performed well, and incorporated captured 105s into the Atlantic Wall defences and the arsenals of various occupation forces.

During the 1930s, the leaders of the French Army belatedly came to the realization that the need existed to modernize their field artillery. Two

French Retreat

French soldiers man a Canon de 105 mle 1935 B field gun during the desperate attempt to stem the German tide in the spring of 1940. Note the 'toed in' wheels which provide extra protection for the crew.

weapons were produced in response. The first of these was the Schneider 105mm M1934 S, which actually seemed to have more characteristics of a howitzer. Schneider received only limited orders for the weapon due to the anticipated availability of the 105mm M1935 B, which was developed by the Atelier Bourges, a state-sponsored design firm. The M1935 was an advanced system with a split trail, similar to American carriages, for added stability. It fired a 15.7kg (34.6lb) shell a distance of 10,305m (11,270yds). Although more than 600 M1935s were ordered, only about 230 of these and scarcely 140 of the M1934 were in service in 1940. The French had given priority to anti-tank guns.

Reliance on fixed fortifications

The French also placed great reliance on the fixed fortifications of the Maginot Line, a system of heavy fortresses constructed between 1930 and 1940 and named for the French minister of defence, André Maginot. Comprising more than 500 separate buildings, which housed up to 1000 soldiers each, the Maginot Line extended from Luxembourg to the Swiss frontier and was intended to serve as a barrier to German invasion from the east. In particular, it was to protect the long-disputed provinces of Alsace and Lorraine and provide the French with precious time to mobilize in the event of war. It had infantry casemates armed with machine guns, six forward fortresses known as ouvrages, and turrets for observation and housing artillery as large as 135mm guns.

Maginot explained to the press:

'We could hardly dream of building a kind of Great Wall of France, which would in any case be far too costly. Instead we have foreseen powerful but flexible means of organizing defence, based on

The fortifications of the French Maginot Line provided little opposition to the German offensive in 1940. Rather than engaging the forts, the Germans simply bypassed them.

Germans, but these are in the minority. Most observers conclude that the Maginot Line was a fantastic anachronism and the product of a French failure to come to terms with the reality of modern war. Mobile firepower, exemplified by the early successes of the German blitzkrieg, rendered the string of fortifications obsolete.

Red star rising

During the 1930s, the Soviet Union arguably possessed the largest complement of artillery among the armies of Europe. The Red Army cadre suffered through the purges of a paranoid Joseph Stalin during the 1930s and endured the costly Winter War with Finland at the end of the decade. Throughout, the Soviet design bureaus maintained a conservative approach to artillery technology.

Early modernization efforts concentrated on an improvement to the 76.2mm field gun M00/02 in 1930, which included updated ammunition and the placement of new barrels on some weapons, resulting in their designation as the M02/30. Six years later, the 76.2mm M1936 field gun, which placed the 76.2mm cannon on the carriage of the 107mm field gun, was introduced. Designed with a wide angle of traverse, the new configuration was in keeping with the premise that every gun in the Soviet arsenal should possess a degree of anti-tank capability. The 76.2mm M1936 performed slightly better than the 02/30, firing its 6.4kg (14.1lb) shell a maximum range of 13,848m (15,145yds). As the Wehrmacht rapidly advanced through Russia in the summer of 1941, a great number of the M1936

the dual principle of taking full advantage of the terrain and establishing a continuous line of fire everywhere.'

In the event, when the German attack came on 10 May 1940, the primary thrust of their army only brushed the Maginot Line as it swept through the Low Countries and the Ardennes Forest. Within two weeks, the Germans had reached the Channel.

Some historians reason that the Maginot Line did indeed serve as a deterrent to direct assault by the

152MM HOWITZER M1937

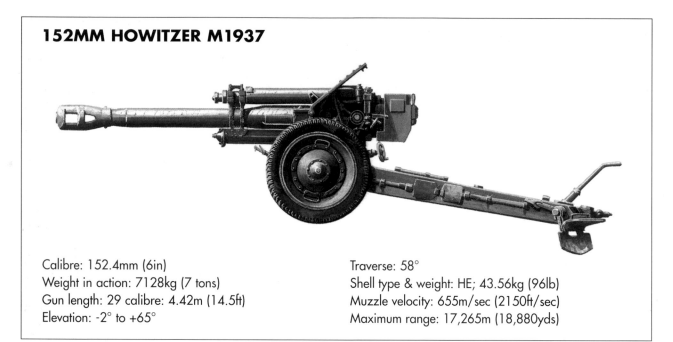

Calibre: 152.4mm (6in)
Weight in action: 7128kg (7 tons)
Gun length: 29 calibre: 4.42m (14.5ft)
Elevation: -2° to +65°

Traverse: 58°
Shell type & weight: HE; 43.56kg (96lb)
Muzzle velocity: 655m/sec (2150ft/sec)
Maximum range: 17,265m (18,880yds)

guns were captured by the Germans, redesignated the 76.2mm FK 296(r) or the 76.2mm Pak 36(r) and turned against their former owners. Another variant, with a shorter barrel, had been introduced in 1939.

As a great deal of Soviet territory was occupied by the Germans, the challenge of maintaining updated artillery was met through the ingenious employment of available manufacturing capacity. The 76.2mm field gun M1942 was the marriage of the M1939 gun and a new split carriage. The M1942 holds the distinction of being produced in the largest numbers of any field artillery piece of World War II. Out of necessity, it was simple to produce and to operate, using the same type of ammunition as that of the magnificent T-34 medium tank and numerous other guns. Firing a

6.2kg (13.7lb) shell, the M1942 maximum range was 13,213m (14,450yds) with an elevation of 37°. Reportedly, the weapon is still in use in some armies today, and the sheer weight of its numbers eventually contributed to the turning of the tide on the Eastern Front.

Additional Soviet artillery pieces of note include the 152mm guns and howitzers, and the 203mm howitzer M1931. Although the origins of the 152mm weapons predate World War I, the Red Army updated both the weapons in 1930. Seven years later, a hybrid piece proved versatile, as did another similar weapon, the 152mm Gun Howitzer M1910/34. The Model 1937 and 1910/34 constituted the core of the Red Army's heavy field artillery throughout the war. Captured pieces were highly prized by the Germans and placed into

service on both the Eastern and Western fronts. The Model 1937 fired a 43.5kg (95.9lb) shell a maximum range of 17,264m (18,880yds). The 203mm howitzer M1931 was the heaviest calibre field weapon deployed by the Soviets in World War II. Mounted on the tracks of a converted tractor, it hurled a heavy shell of 100kg (220.4lb) a distance of 18,025m (19,712yds).

Gotterdammerung at Stalingrad

The great industrial city on the Volga bore the name of Soviet premier Joseph Stalin. The mighty struggle for its possession raged from August 1942 until the surrender of the encircled German Sixth Army in February 1943. Once German forces entered Stalingrad, Adolf Hitler refused to allow them to withdraw, even after 3500 Red Army guns had devastated Rumanian army units and the jaws of a two-pronged Soviet offensive slammed shut, trapping 300,000 soldiers. Eventually, the Germans lost nearly 150,000 dead and 91,000 taken prisoner.

In Enemy Hands

During their early victories on the Eastern Front, the German war machine captured thousands of Soviet artillery pieces and other weapons abandoned by the retreating Red Army. Along with these spoils of war, millions of rounds of shells and unused ordnance were also taken. In many cases, the Germans found the Soviet artillery to be equal to or better than their own. Often, the Soviet guns were placed in service with the Wehrmacht. Some of the Red Army weapons were actually rebored to accept German ammunition and were subsequently turned on the forces they were intended to serve.

Red Army newspaper correspondent Vasili Grossman wrote of the fighting around Stalingrad:

'Our artillery played an invaluable role in repulsing the German attacks. Fugenfirov, the commander of the artillery regiment, and the battalion and battery commanders were up in front with the Siberian Division. They were in direct radio contact with the firing positions, and the crews of dozens of powerful long-range guns across the Volga breathed as one with the infantry, sharing their every joy and sorrow, their every anxiety. The artillery was invaluable in dozens of ways: it covered the infantry positions with a solid shield of fire, mangled the German tanks as if they were made of cardboard, those heavy tanks that the anti-tank units were unable to deal with, mowed down the enemy soldiers advancing under cover of the tanks, pounded now a square, now an enemy troop concentrations, blowing up ammunition dumps and sending mortar batteries sky-high. At no other time in the course of the war had the infantry felt such friendly support from the artillery as at Stalingrad.'

The ferocity of the concentrated Red Army artillery bombardment was captured in a letter written by German soldier Heinz Küchler:

'The Russians began a bombardment of unprecedented ferocity. Everything became opaque, and the sun vanished from our eyes… Screams of fear froze in our constricted throats… Suddenly a human figure crashed into our hole …[and] shouted to us: "My whole company was wiped out!" He carefully lifted his head just over the edge of the

During the battle for Stalingrad, a Red Army soldier waves while the crew of a heavy mortar demonstrates the technique for loading and firing the weapon.

embankment as a series of explosions began to rip through the air beside us. His helmet and a piece of his head were sent flying, and he fell backward, with a horrifying cry. His shattered skull crashed into Hals's hands, and we were splattered with blood and fragments of flesh. Hals threw the revolting cadaver as far as he could, and buried his face in the dirt. Nothing remains for those who have survived such an experience but a sense of uncontrollable imbalance, and a sharp, sordid anguish… We felt

like lost souls, who had forgotten that men are made for something else.'

Following the terrible losses sustained during the German offensive in the summer of 1941, the Soviet high command reorganized its artillery. Previously, artillery units had been incorporated within the rifle divisions of the Red Army. By the time of the great battle at Stalingrad, artillery formations as large as army corps size were present within the Soviet order of battle. Throughout the

HEAD TO HEAD: *76.2mm Model 1942* VERSUS

The Soviet-made 76.2mm Field Gun Model 1942 was produced in greater numbers than any other artillery piece of World War II. Its basic design allowed for rapid production and solid performance for the duration of the Great Patriotic War. The reliable weapon remains in service to this day.

76.2mm Model 1942

Calibre: 76.2mm (3in)
Weight in action: 1116kg (2460lb)
Gun length: 3.24m (10ft 7in)
Elevation: -5° to +37°
Traverse: 54°
Shell type & weight: AP/HE; 7.54kg (16.62lb)
Muzzle velocity: 740m/sec (2427ft/sec)
Maximum range: 2000m (2190yds)

STRENGTHS

• Excellent range
• Ease of production
• Available in great numbers

WEAKNESSES

• Little self-propelled capability
• Conservative design
• Limited ordnance options

105mm leFH 18

Based on a previously successful design of World War I and produced by Rheinmetall, the 105mm leFH 18 debuted in 1935 as a stalwart field howitzer with the Germany Army. Compared to previous designs of lighter calibre, the leFH 18 offered better firepower with little or no sacrifice in mobility.

105mm leFH 18

Calibre: 105mm (4.13in)
Weight in action: 1985kg (4376lb)
Gun length: 24.8 calibre: 2.61m (102.75in)
Elevation: -6.5° to +40.5°
Traverse: 56°
Shell type & weight: HE; 14.8kg (32.65lb)
Muzzle velocity: 470m/sec (1542ft/sec)
Maximum range: 10,675m (11,675yds)

STRENGTHS

• Excellent firepower
• Good range
• Heavy projectile

WEAKNESSES

• Excessive weight
• Bulky carriage
• Lack of general mobility

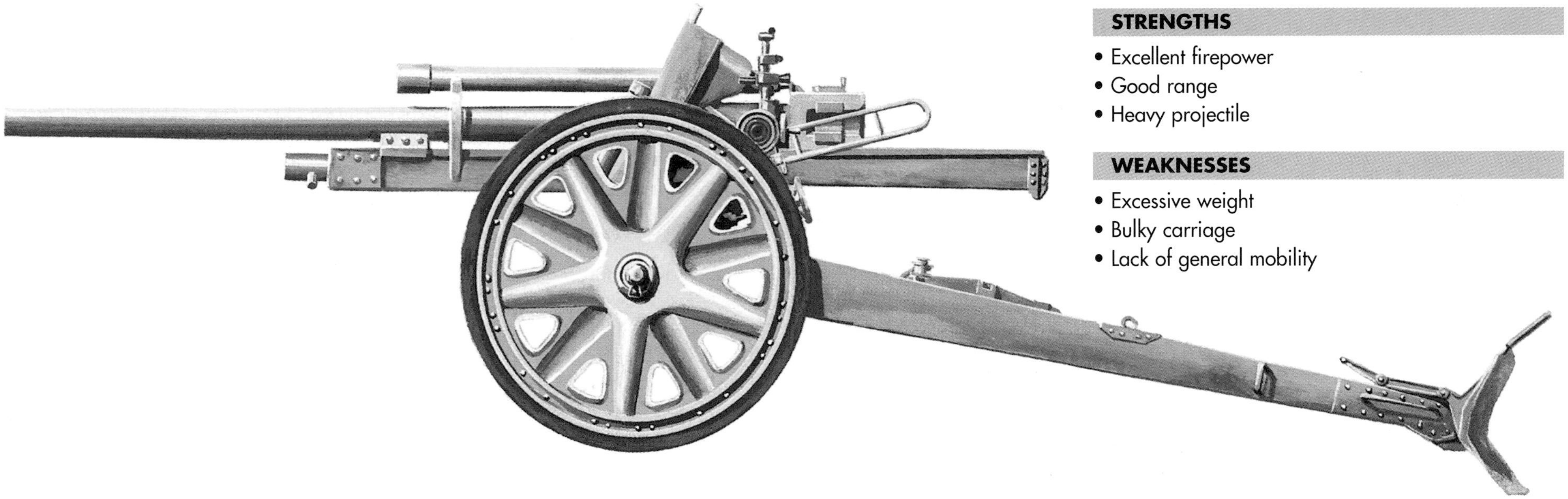

war, the Red Army relied on the maxim that the rapid deployment of artillery could decide the course of a battle. At times, Red Army artillery was concentrated to a strength approaching 500 pieces per mile of front.

Treaty and tragedy

The terms of the Treaty of Versailles placed blame for World War I squarely on the shoulders of the German people. Intended to be harsh, the treaty was dictated to Germany, which was a wholly unexpected turn of events following their agreement to end hostilities. France sought a secure border with its traditional adversary, while Great Britain and the United States wanted to prevent the future rise of militarism in Germany but at the same time allow the nation to maintain a defensive posture against the spread of Communism.

Militarily, the German Army was reduced in strength to 100,000 personnel. Tanks, an air force and submarines were forbidden. Only six battleships were permitted. Territory west of the Rhineland and to the east of the Rhine River was designated a demilitarized zone, and an Allied army was destined to occupy the west bank of the great waterway for 15 years. Germany was required to cede territory to France, Belgium, Denmark, Czechoslovakia and Poland, while the Saar, Memel and the port city of Danzig were to be under the administration of the League of Nations. Germany was also saddled with financial responsibility for the war and forced to agree to pay astronomical reparations.

An audience shouts 'Sieg heil!' and gives the Nazi salute in response to a speech by Adolf Hitler delivered when the Führer had reached the zenith of his power.

The German people were particularly incensed by the idea of bearing guilt for the war. The diplomats who signed the treaty were referred to as the 'November Criminals'. In reality, they had no recourse. German foreign minister Ulrich Graf von Brockdorff-Rantzau petitioned:

'The German Delegation again raise their demand for a neutral inquiry into the question of responsibility for the war and of guilt during the war. An impartial commission should have the right of inspecting the archives of all belligerent countries and examining, as in a court of law, all chief actors of the war… The high aims which our adversaries were the first to establish for their warfare, the new era of a just and durable Peace, demand a Treaty of a different mind. Only a cooperation of all nations, a cooperation of hands and intellects, can bring about a permanent peace.'

The words of Brockdorff-Rantzau proved prophetic. Conditions would become ripe for the rise of the Nazi Party in Germany. Adolf Hitler promised a redress of the injustices done to the German people in the hated treaty. On 1 January 1933, Hitler became chancellor of Germany, and charted a course of rearmament that undermined the terms of the Treaty of Versailles, and plunged the world into war again with the invasion of Poland on 1 September 1939. The Führer had even concluded arrangements with the Soviets for cooperation in arms production and military training. Eventually, Nazi Germany and the Soviet Union signed a non-aggression pact, which Hitler subsequently ignored with the launching of Operation Barbarossa, the invasion of Russia, on 22 June 1941.

The dreaded 88

Erwin Rommel, the famed Desert Fox and commander of the Afrika Korps, stated bluntly, 'The struggle in the desert is best compared to a battle at sea. Whoever has the weapons with the greatest range has the longest arm. The longest arm has the advantage. We have it in the 88mm gun.'

Originally designed as an anti-aircraft weapon, the 88mm gun gained lasting notoriety in an anti-tank role. Restricted to only certain types of artillery production by the Treaty of Versailles, a research and development team from Krupp embarked for Sweden and worked initially on a 75mm flak gun in cooperation with Swedish arms

manufacturer Bofors. When the German Army requested a heavier weapon, the result was the 88mm Flak 18. As an anti-aircraft weapon, the Flak 18 was immediately considered successful, having debuted in 1933 and seen limited service during the Spanish Civil War. The 88mm shell weighed 9.2kg (20.3lb) and reached an altitude of 8000m (26,245ft) with a muzzle velocity of 820m (2690ft) per second. The Flak 18 was later improved with a multi-piece barrel rather than a single piece barrel, which allowed only the worn section to be replaced when necessary. This revision was designated the Flak 36. The Flak 37, with improved fire-control data transmission, soon

followed. The weapon was serviced by a crew of nine, and when used in an anti-tank role, its automatic spring mechanism, which ejected a spent casing and thrust a new shell into the breech automatically, was often discarded.

Another 88mm gun, the Flak 41, was developed by the firm of Rheinmetall, both as an improved anti-aircraft and anti-tank weapon. However, the design was plagued with problems and not implemented in large quantities. By the time the war ended, the Germans had only placed slightly more than 300 in service. Although the Flak 41 did have issues, it was an effective anti-aircraft weapon, firing the 88mm shell to a ceiling of 14,700m

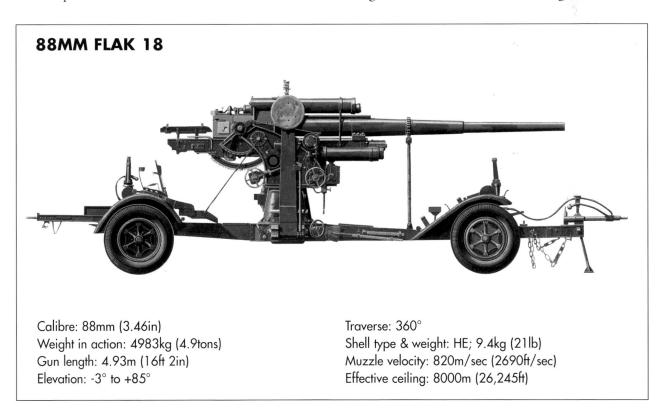

88MM FLAK 18

Calibre: 88mm (3.46in)
Weight in action: 4983kg (4.9tons)
Gun length: 4.93m (16ft 2in)
Elevation: -3° to +85°

Traverse: 360°
Shell type & weight: HE; 9.4kg (21lb)
Muzzle velocity: 820m/sec (2690ft/sec)
Effective ceiling: 8000m (26,245ft)

Advancing German soldiers pull an 88mm Flak 18 multi-purpose gun across a pontoon bridge that engineers have constructed over a river in Western Europe.

(48,230ft) and easily coping with high-flying formations of Allied bombers. The Flak 41 is not to be confused with the more notable 88mm Flak 18, 36 and 37 series.

One British tank commander, whose armoured vehicle had been disabled by an 88mm round, was surprised by the size of the gun after he had been taken prisoner. 'It doesn't look like much, but nothing can be done against it,' he said. Another

remarked that it was decidedly unfair to use an anti-aircraft gun against tanks.

Colin Smith describes one encounter between British tanks and the powerful 88s during an operation at El Alamein to secure an objective codenamed Snipe:

'At first this worked out quite well. Turner's anti-tank guns picked off some of the panzers that had turned to face this new threat. The machine-gunners in the Shermans – new to the desert and unburdened by notions of chivalry between opposing tank crews – shot the survivors. But when 24 Brigade got onto Snipe itself, the high-

turreted Shermans came under devastating anti-tank fire from the German 88s. Soon, seven British tanks were ablaze and as Turner's riflemen risked their lives trying to rescue their crews it became apparent that Snipe was no place for any target that could not be concealed below its low natural parapet.

'The Shermans accordingly withdrew east of Kidney Ridge, pursued by the taunts of the Germans who got onto their radio frequency and proceeded to demonstrate that they knew an alarming amount about their opponents. The 15th Panzers must have had an energetic propaganda team at divisional headquarters for Flatow's [a British tank commander] radio began picking up a very Germanic rendering of English North Country voices, spreading alarm and despondency. "Aye," said one voice, "it's the 45th, 41st and 47th regiments, they come from Lancashire and Yorkshire. We'd be much better off at home in our gardens with our wives... We can't do anything against the German artillery... These 88mms are so accurate... I don't know what we're fighting for."'

'It was all in that strain,' Flatow later recalled, 'two soldiers talking to each other... Believe me, it was incredibly demoralizing. I switched off my set so my crew couldn't hear it – as it was, they were rather windy.'

Psychological warfare

The almost mythical status achieved by the 88mm gun is evidenced by the fact that in September 1942, the Military Intelligence Service of the US Army published information on the 88mm weapon in its periodical intelligence bulletin.

'Field Marshal Erwin Rommel's use of the 88-mm. anti-aircraft gun as an offensive anti-tank

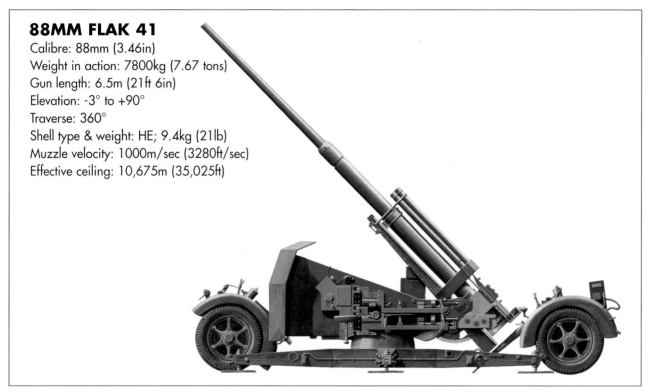

88MM FLAK 41
Calibre: 88mm (3.46in)
Weight in action: 7800kg (7.67 tons)
Gun length: 6.5m (21ft 6in)
Elevation: -3° to +90°
Traverse: 360°
Shell type & weight: HE; 9.4kg (21lb)
Muzzle velocity: 1000m/sec (3280ft/sec)
Effective ceiling: 10,675m (35,025ft)

withdrawn for offensive operations somewhere else, and the 88-mm begins its mission of trying to pierce the armor of approaching hostile tanks.

'What are the gun's weaknesses? First, it makes a good target for dive bombers, even though one of its duties is to oppose aircraft. The 88 has a hard time placing its fire on dive bombers. They can come down at a furious speed, blast the 88, and get away successfully. But the really important weakness of this, or any other gun, lies not in its manufacture but in its crew. The gun is not alive; the men are. They know that it may not be aircraft or long-range artillery which will end the big gun's usefulness to the Axis, but a detachment of perhaps 20 American soldiers. To meet such a threat, the crew wear their rifles strapped to their backs in readiness for close combat…'

Thunder of the Third Reich

The shattered German economy was a contributing factor to the rise of the Nazis during the years between the wars. When Hitler consolidated power in 1933, he did so with the knowledge that both Krupp and Rheinmetall, the country's two major arms manufacturers, remained viable. In the wake of World War I, foreign and domestic markets for weapons had essentially evaporated. Therefore, a period of research and development of new artillery types ensued during the 1920s.

Given the high level of quality of submissions by both companies in efforts to win the contract from the new government for a 150mm field howitzer, an innovative compromise was reached. The Rheinmetall gun was placed on a carriage designed by Krupp, and the 150mm (actually 149mm) Field Howitzer 18 emerged and became the stalwart among heavier German field artillery of World War

weapon in Libya has caused so much discussion among American soldiers everywhere that this seems a good time to describe it. In plain English, there is nothing strange or unusual about the 88. To the question "Is it vulnerable?" the answer is "Yes!"

'Back in the 10-year period before Hitler came into power, the German 88-mm anti-aircraft gun was designed and built in secret. In those days the German Army was rigidly limited as to men and materiel. It is known that the designers of the gun were chiefly interested in constructing a double-purpose anti-aircraft and anti-tank weapon. The news of the gun's anti-tank capabilities was not

allowed to leak out, however, and not until the Nazis invaded Poland did the world discover what the German designers had perfected.

'Basically, the 88-mm. is a tractor-drawn gun for firing on moving targets… It has a rate of fire of 25 rounds per minute, or slightly better, and is capable not only of a great volume of fire, but of extreme accuracy against moving targets of any type. This applies to targets on the ground as well as those in the air. When the 88-mm. is to attack armored vehicles, it is provided with a special armor-piercing projectile.

'Rommel generally sends the gun into position under cover of medium tanks. The tanks are then

HEAD TO HEAD: *Ordnance, Q.F. 25 pdr Mk 2* VERSUS

One of the most famous artillery pieces of World War II, the 25-pounder was designed between the wars to provide the British Army with a versatile weapon that could fulfil the role of both a gun and a howitzer. The weapon took centre stage during the heavy bombardment that preceded the offensive at El Alamein.

Ordnance Q.F., 25 pdr

Calibre: 87.6mm (3.45in)
Weight in action: 1800kg (3968lb)
Length: 2.4m (7ft 11in)
Elevation: -5° to +40°
Traverse: on carriage 8°; on firing table 360°
Shell type & weight: 11.34kg (25lb)
Muzzle velocity: 532m/sec (1745ft/sec)
Maximum range: 12,253m (13,400yds)

STRENGTHS

• Versatility as gun and howitzer
• Innovative carriage design
• Good mobility

WEAKNESSES

• Limited anti-tank ammunition
• Scarcity early in World War II
• Tractor transport

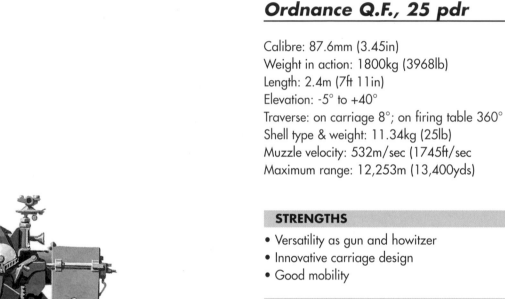

88mm Pak 43/41

On the eve of World War II it was apparent to German armament designers that improved aircraft performance would necessitate the development of a new generation of anti-aircraft cannon. The 88mm Pak 43/41 was the result. Originally debuted in 1941, it took two more years for mechanical problems to be resolved.

88mm Pak 43/41

Calibre: 88mm (3.46in)
Weight in action: 4380kg (9656lb)
Barrel length: 6.61m (21.69ft)
Elevation: -5° to +38°
Traverse: 56°
Projectile type & weight: APCR: 7.3kg (16lb); AP/HE: 10.4kg 23lb)
Muzzle velocity: APCR: 1130m/sec (3707ft/sec); AP/HE:1000m/sec (3280ft/sec)
Maximum range: 4000m (4375yds)

STRENGTHS

• High rate of fire
• Advanced technology
• Effective ceiling

WEAKNESSES

• Mechanical malfunctions
• Limited availability
• Production delays

II. The weapon was produced in large numbers, originally configured to be drawn by horses and later modified for towing behind a halftrack or truck. When large numbers of the 152mm Soviet field gun fell into German hands, it was discovered that the Wehrmacht weapon was outranged by its Soviet counterpart. Modifications to the propellant charge and the addition of a muzzle brake proved less than adequate and were abandoned. Later in the war, a self-propelled variant was produced. The 150mm field howitzer delivered a 43.5kg (95.9lb) shell a maximum range of 13,323m (14,570yds) with an elevation of up to 45°.

A pair of 150mm (actually 149.1mm) field guns were used by the German Army in World War II. Rheinmetall introduced its Model 18 in 1938. Although the military commanders were pleased with both the long range of the gun, some 24,506m (26,800yds), and the 43kg (94.8lb) projectile it fired, the transportation of the weapon proved to be a major drawback. The long barrel of the Model 18 required that the barrel be detached from the carriage and moved separately if any great distance was to be covered. To complicate matters, the low rate of fire – only two rounds per minute – placed the gun at a decided disadvantage. When its deficiencies were noted in the field, production of the Model 18 ceased and most examples were relegated to coastal defences.

Intended as both a functional field and coastal defence weapon, the Krupp-designed 150mm (actually 149.1) cannon was produced for delivery to Turkey on the eve of the outbreak of war. When hostilities prevented further deliveries to Turkey, the weapon was adopted by the German Army. For transportation, the 150mm cannon had to be broken down into two pieces, three in its coastal

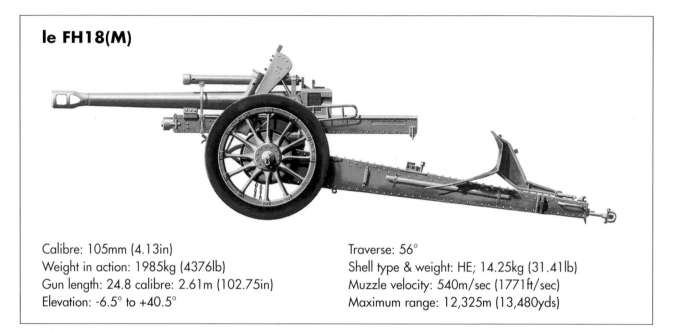

le FH18(M)

Calibre: 105mm (4.13in)
Weight in action: 1985kg (4376lb)
Gun length: 24.8 calibre: 2.61m (102.75in)
Elevation: -6.5° to +40.5°

Traverse: 56°
Shell type & weight: HE; 14.25kg (31.41lb)
Muzzle velocity: 540m/sec (1771ft/sec)
Maximum range: 12,325m (13,480yds)

configuration. Eventually, due to difficulties with ammunition, the weapon was utilized in a training role. Only about 40, firing a 43kg (94.8lb) projectile a maximum distance of just over 9144m (10,000yds), were produced.

One of the great Krupp innovations of World War II was the double recoil carriage, which incorporated rails in the carriage to complement the standard recoil system located close to a weapon's barrel, therefore absorbing virtually the entire shock of a heavy gun's recoil and dramatically improving accuracy. The heavy 210mm (actually 210.9mm) Mörser 18 fired a 121kg (266.8lb) shell a range of 16,706m (18,270yds). Subsequently, however, it was discovered that a smaller weapon, the 170mm (actually 172.5mm) 18, was nearly as effective as the 210mm Mörser 18, while also offering a greater range of 29,599m (32,370yds) firing a 68kg

(149.9lb) projectile. With the double recoil carriage employed, both weapons were used until the end of the war.

The massive 355mm (actually 356.6mm) Howitzer M.1 weighed a staggering 78,000kg (171,960lb) and fired a shell weighing as much as 926kg (2041.5lb) up to 20,848m (22,800yds). The howitzer was designed by Rheinmetall in 1935 and was placed in service in 1939 with the double recoil carriage as standard. The ponderous weapon required dismantling into six pieces for transportation, and an electrically operated gantry system was necessary to assemble and disassemble the howitzer. Records of its service in the field are sketchy, and the number of 355mm M.1 Howitzers produced is a matter of conjecture. Only one unit, the 1st Motorized Artillery Battery 641, is known to have operated the weapon.

The 355mm Howitzer M.1 was originally designed as a heavier replacement for the 240mm Cannon 3, of which less than a dozen are known to have been produced and served by crewmen of Motorized Artillery Battalion 83. This weapon was also broken down into six components for transport, but a number of winches and other innovations were incorporated to assist with more rapid assembly. Oddly enough, the Rheinmetall design was actually produced in Essen at the Krupp manufacturing facility. When Krupp engineers inspected their competitor's design, they determined to improve it and unveiled the 240mm Cannon 4. An Allied air raid put an end to an ambitious project involving a self-propelled design, when bombs destroyed the prototype in 1943. The 240mm Cannon 3 fired a 1527kg (3,366.5lb) shell a maximum range of 37,500m (41,010yds).

In 1935, the German Army asked for a standard 105mm field howitzer that could supply a heavier round than the earlier 77mm weapon used during World War I. The requirement was met by Rheinmetall, with its 105mm leFH 18. This was a standard export to countries of Eastern Europe, Spain and South America, but when war came, the need for additional range resulted in a revision with a muzzle brake and additional propellant charge. This version, introduced prior to the campaign against Russia, was designated the 105mm leFH 18(M). The major shortcoming of the 105mm leFH 18 was its heavy weight of 1955kg (4310lb). The weapon frequently became bogged down, although it was intended to be moved by a tracked vehicle. To address this issue, the carriage of the 75mm Pak 40 anti-tank weapon was added later. The newly named 105mm leFH 18/40 proved to be only slightly lighter than its predecessor. The standard weapon fired a 14.8kg (32.6lb) shell a distance of up to 12,324m (13,478yds).

Yet another marriage of convenience between a Rheinmetall gun and a Krupp carriage occurred with the 105mm Cannon 18 and Cannon 18/40, which reached field units of the interwar German Army in 1934, a full four years after both armaments firms had developed prototypes for a new long-range fire support weapon. The Cannon 18 was temporarily adopted as standard; however, it was soon discovered that the 150mm pieces available provided bigger bang for the buck without adding appreciably to the weight load in transport.

Initially, the 105mm Cannon 18 was moved by a team of horses, but to do so the barrel and carriage were separated. By the time tracked transport became available, the enthusiasm for the weapon had waned, and it was a low production priority. The 105mm Cannon 18/40 (sometimes also referred to as the 105mm sK 42) was a longer range version of the Cannon 18 with an extended barrel, and was produced in limited quantities. A number

A battery of German field cannon is serviced by artillerists in preparation for the support of a coming attack. Early in World War II, German artillery was unsurpassed.

of the 105mm Cannon 18 family of weapons eventually defended positions along the Atlantic Wall. The gun was capable of firing a 15kg (33.1lb) shell a maximum range of 19.074m (20,860yds).

Between the wars, the Germans abandoned their original 77mm field cannon for a 75mm design, which had become something of a worldwide standard. The old 77mm Field Cannon 16, of World War I vintage, was rebarrelled to 75mm, and the new weapon, designated the 75mm FK 16 nA, reached the army in 1934. Although the 75mm FK 16 nA was already outdated to a large extent, it served until the end of World War II, was used in a training role, and actually achieved some degree of notoriety when one gun was reported to have destroyed at least 10 Allied tanks in Normandy before being silenced.

A lighter field weapon, the 75mm leFK 18, was initially intended for cavalry use. It was designed by Krupp in 1930 but not placed in service until eight years later. It was produced only in limited numbers as the focus of military planners shifted to the heavier 105mm weapons. The 75mm Field Cannon 16 fired a 5.8kg (12.8lb) projectile a maximum range of 12,875m (14,080yds), while the leFK 18 used the same ammunition with a range of 9427km (10,310yds).

Skoda in German hands

When Nazi Germany exerted international pressure on Czechoslovakia in 1938–39, one of Hitler's ultimate objectives was to make use of the massive Skoda armaments operation at Pilsen. Czechoslovakia had been constituted as a nation following the end of

A German crewman adjusts the setting of a 150mm field howitzer, which was a mainstay of the Wehrmacht and served on all fronts throughout World War II.

World War I and the dissolution of the Empire of Austria-Hungary. During the 1920s and 30s, Skoda continued to produce weapons for export to countries around the world, upgrading existing designs already proven in combat and introducing new weapons as well. With its history of quality armaments production, the Skoda name was, perhaps, second only to that of the venerable Krupp firm in prominence, and the company was eventually required to turn out not only artillery but also armoured vehicles for the Nazi war machine.

The Skoda 149mm vz 37 Howitzer (K1) debuted in 1933, and the entire production run was exported to Turkey, Yugoslavia and Rumania. The Czech Army declined to order the K1 but requested further research into a weapon that met more exacting specifications. The result was the K4, with a shorter barrel than its predecessor and an easier horse-drawn

British soldiers inspect a captured German 150mm field gun. Production of the 150mm cannon was discontinued in favour of heavier weapons as World War II progressed.

transport capability. It fired a 42kg (92.6lb) shell up to 15,101m (16,515yds), but only a few of the K4 were actually delivered to the Czech Army before the country was essentially surrendered at the Munich Conference in the autumn of 1938. The Germans took possession of the Skoda Works and placed the capable K4 in service as the 150mm Heavy Field Howitzer 37(t). The (t) denoted that the weapon was of Czech origin.

Other artillery pieces produced by Skoda during the interwar years were pressed into service with the Germans, including the 220mm Howitzer, which fired a 128kg (282.2lb) shell a range of 14,200m (15,530yds). The 220mm Howitzer of the period was a response to the need for a more mobile heavy calibre weapon than had been available during World War I, and a number of these were sold to Yugoslavia and Poland with the additional designation of M.28. The large howitzer was primarily used by the Germans as a coastal defence weapon and during the siege of Sevastopol in the East.

The 76.5mm Cannon vz 30 and 100mm Howitzer vz 30 were Skoda weapons developed in the late 1920s to be used as a field artillery and anti-aircraft weapon and a mountain gun or field howitzer respectively. Although it was unremarkable during performance in the anti-aircraft role, the 76.5mm was adopted in 1930 by the Czech Army,

Right: During a training exercise, soldiers of the Indian Legion, disillusioned prisoners of war recruited to serve with the German Army, prepare to fire a Model 16 field gun.

later in Albania and Greece. Initially suspicious of Hitler and the Nazis, Mussolini later embraced them as political and military allies. In fact, it was the Germans who eventually came to the rescue of the Italian Army, which was roughly handled on multiple fronts during the dictator's misadventures.

The effort to realize the Fascist dream of a new Roman Empire and a Mediterranean Sea that was essentially under Italian control was hampered from the beginning by a lack of the natural resources necessary for a modern army to prosecute an effective, sustained military campaign. Sufficient quantities of oil, steel, coal and other resources were rarely available, and at times the Italian forces in the field found their weapons markedly inferior to those of the Allies.

Skoda Lineage

The circumstances of war placed the valuable Skoda armaments production facilities, headquartered in Pilsen, Czechoslovakia, at the disposal of Hitler and the Third Reich. Founded in 1859 by Count Wallenstein-Varterberk, the enterprise was initially a manufacturer of steel and machinery. Producing equipment of all types, including steam engines, boilers, and bridge and railway components, the company was acquired in 1869 by Emil Skoda, a successful engineer and businessman. Shortly thereafter, the company name was changed to reflect this ownership. The company grew steadily and became the chief supplier of artillery and ammunition for the army of Austria-Hungary during World War I. By 1917, the Skoda works in Pilsen alone employed 35,000 people. In 1939, following the German occupation of Czechoslovakia, Skoda became a primary supplier of armaments to the Nazis.

as was the 100mm vz 30. Few of either model reached Czech service, and both were used during World War II by the Germans. The 76.5mm fired an 8kg (17.6) shell at a range of up to 13,231m (14,770yds), while the 100mm vz 30 fired a 16kg (35.3lb) projectile a maximum distance of 16,002m (17,500yds).

Improving on a 1914 design, the Skoda 100mm Howitzer vz 14/19 had a longer barrel and higher performance ammunition than its predecessor, the 100mm vz 14. By 1939, large numbers of the 14/19 were in service across Europe, and the German Army employed these Czech manufactured guns during the campaign in France in 1940. The 14/19

When Germany annexed Czechoslovakia, the Skoda arms works began supplying weapons to the Wehrmacht. Here, a German crew fires a Skoda 149mm vz 37 gun.

fired a 14kg (30.9lb) shell a maximum distance of 9973m (10,907yds).

Ill-fated dream

When Benito Mussolini and the Fascists ascended to power in Italy in 1922, the totalitarian leader initiated a program of military spending. His intention was firstly to modernize the nation's armed forces and, secondly, to prepare for campaigns of territorial conquest in Africa, and

Visiting women talk with artillery troops in the shadow of a gigantic Skoda 220mm howitzer, which was designed to provide plunging fire against hardened targets.

was powerful enough to be used in an infantry support or anti-tank role. The army clamoured for more of the weapon throughout the war, and it was used by the Germans extensively after Mussolini was deposed and Italy changed sides to fight with the Allies in 1943. The major deficiency of the 75/32 was that there were never enough of them.

Another Italian field artillery piece worthy of note is the Obice 75/18 Model 35, which was an adapted version of a successful mountain howitzer, the Obice 75/18 Model 34. Few of the weapon were ever placed into service, and its major drawback was that it could not be disassembled. Its range was an

Adolf Hitler extends the Nazi salute as troops of the Wehrmacht pass in review during a major parade. To Hitler's right is Italian fascist dictator Benito Mussolini.

Following World War I, the basis of the Italian Army's artillery complement often depended upon outdated weapons of the Great War. When the Fascists embarked upon their program of upgraded armaments, however, the artillery was given a lower priority than the air force or the navy. Even so, the Cannon 75/32 Model 37, with its 75mm gun and range of 12,504m (13,675yds) firing a 6.3kg (13.9lb) shell, compared favourably to the field artillery of other nations. Originally designed for transport by a tracked or wheeled vehicle, the 75/32

impressive 9565m (10,460yds) firing a standard 6.4kg (14.1lb) shell.

A prototype of the Obice 210/22 Model 35 – a 210mm howitzer firing a 133kg (293.2lb) shell a maximum range of 15,408m (16,850yds) – was introduced by the Ansaldo manufacturing firm in 1935. Not until three years later did the Italian Army order 346 of the big weapon. Needless to say, supply problems hampered delivery. Ironically, a number of the 210/22 howitzers produced were sold to Hungary and eventually placed in service on the Eastern Front. Fewer than 25 of them had even been completed by the autumn of 1942. Those that did enter service proved to be reliable and effective weapons. Transport required that the howitzer be broken down into at least two components, and for more rigorous movement it could be further disassembled into four parts. When German forces took possession of the few 210/22 weapons available, they redesignated it the 210mm Howitzer 520(i).

Clash in the desert

The history of the British 7th Armoured Division acknowledged the prowess of Italian artillery encountered during operations in the Libyan desert, but further noted the consequences of inadequate supply, which limited its overall effectiveness:

'The A9 Cruisers went ahead for the Italian square, while the RHA Anti-Tank troop fired at it from the right flank, and targeted the soft skinned Italian trucks and exposed infantry. It was only after the second trip around the square that the Italian artillery revealed themselves and opened fire on the British armoured vehicles. The Italian artillery had only high explosive ammunition and no armoured piercing ammunition at their disposal, but they

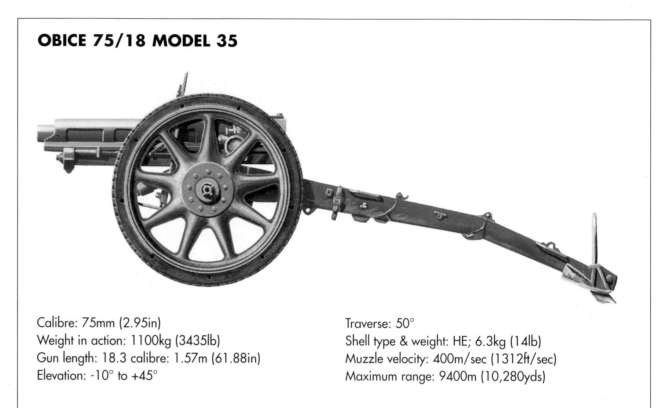

OBICE 75/18 MODEL 35

Calibre: 75mm (2.95in)
Weight in action: 1100kg (3435lb)
Gun length: 18.3 calibre: 1.57m (61.88in)
Elevation: -10° to +45°

Traverse: 50°
Shell type & weight: HE; 6.3kg (14lb)
Muzzle velocity: 400m/sec (1312ft/sec)
Maximum range: 9400m (10,280yds)

were still capable of inflicting damage on the British armoured vehicles. There were no anti-tank guns with the column either, which was a clear violation of the Italian doctrine for combined arms teams operating in conjunction with the infantry and armour. The battle dissolved into four separate fights at each corner of the square with Italian guns and British armour engaging each other, but the exposed Italian gunners soon fell and the infantry broke. There was no lack of courage or skill from the Italian gunners as they engaged the enemy armour, just the lack of armour-piercing ammunition. In the battle, eight more L-3

Tankettes were destroyed by 11th Hussars and nine by 7th Hussars, with all four field guns being destroyed along with over 100 prisoners being taken, with the battle effectively destroying an entire Infantry battalion and two companies of tanks.'

The co-prosperity sphere

Emerging from the Russo-Japanese War as a rival of the West for economic and territorial dominion in Asia, Japan had patterned the organization of its armed forces largely after those of the major European states. Essentially a bystander during

A gun crew services its weapon against a distant enemy position. Acquisition of targets was often based on coordinates provided by forward or aerial observers.

World War I, Japan benefited from territorial concessions and mandates issue by the League of Nations to govern certain territories. The Japanese had also dominated Korea by proxy since 1910.

Faced with shortages of raw materials, which were crucial to industrial growth and to the efficiency of an expanding military, the Japanese were confronted with challenges similar to those of the Italians. However, as militarists came to dominate the government in Tokyo, the decision to expand Japan's territory and claim the resources needed was cloaked in the euphemistic expression of a Greater East Asia Co-Prosperity Sphere. Ostensibly, the Japanese were proclaiming 'Asia for the Asians', but in reality their mantra was 'Asia for the Japanese'. As early as 1931, Japanese troops were fighting in China.

Strangely enough, having experienced the decisive power of artillery at Port Arthur during the Russo-Japanese War and witnessed its awesome power during World War I, military strategists in Tokyo apparently discounted the importance of modern artillery and armored vehicles. One contemporary assessment noted,

'Japanese use of artillery is subject to much criticism. The fundamental fault is that there is generally not enough of it. The weakness in artillery may be the result of lack of appreciation of the need for adequate fire support, or of a feeling that past experience has not demonstrated the need for stronger artillery. The period of daylight fire for adjustment prior to the fire for effect

reduces tactical surprise and diminishes the moral [sic] effect of the preparation. This unwillingness to fire the preparation unobserved at night would suggest low gunnery efficiency. Also, the absence of general support artillery reduces the flexibility of the artillery fires and limits the ability of the division commander to intervene promptly in the action by use of his artillery. From the picture drawn in the tactical problems, one can feel reasonably sure that the Japanese infantry will jump off, even though their extensive preparations have neither destroyed hostile wire nor neutralized the enemy artillery and machine guns.'

Overwhelming force in China

Although Japanese army units relied on antiquated guns, their firepower was considerably superior to that of the defending Chinese. Sketchy records indicate a hodgepodge of 75mm, 100m and 150mm guns in use during the Sino-Japanese War. Even earlier, armed conflict in China had proven to be a one-sided affair.

An article in a January 1933 edition of *Time* magazine reported:

'In the open, Chinese soldiers proverbially run. Cornered, they fight like wild tigers, defend every street corner and doorway, die frothing and screaming defiance... Dog-trotting behind a murderous Japanese barrage the Imperial troops entered Shanhaikwan's dragon-crested South Gate. Firing from cover Chinese riflemen drove them back once, twice. Next artillery battered breaches in the walls, Japanese troops burst through, fought bayonet-to-bayonet with desperate Chinese among the low mud huts of Shanhaikwan's narrow, winding streets. Hurtling from the sky Japanese bombs set the city afire, rained death among soldiers and civilians alike. Japanese gunners, when they finally got the range, concentrated on Shanhaikwan's famed Drum Tower which has sounded warnings for centuries, [and] sent it crashing down in smoke.'

Imperial Japan on the March

Japanese development of artillery lagged significantly behind that of other industrialized nations. Thus, the army of conquest was forced to employ antiquated models during its campaigns on the Asian mainland and elsewhere in the Pacific. A relative few Japanese-designed artillery pieces were actually fielded during the course of World War II. In this photo, advancing Japanese troops are seen from a vantage point adjacent to an antiquated field piece.

While their firepower may have been relatively impressive in confrontations with the Chinese, the Japanese clashed with the Soviets during a series of border incidents in Manchuria in 1939. Superior Soviet artillery took a heavy toll and contributed substantially to the defeat of the Japanese.

Heavily reliant on designs that often predated World War I and were originated by Krupp, the Japanese made at least one notable attempt to upgrade a widely used field cannon. Based on a Krupp weapon dating back to 1905, the Japanese used the 75mm Type 38 throughout the war. They did endeavour to improve the gun with minor recoil system and carriage adjustments, prompting the Allies to give the weapon the designation of 75mm Field Gun Type 38 (Improved). Because of its advanced age, the Type 38 was originally intended for horse-drawn transport, and this configuration never changed. Therefore, the gun was easily rendered immobile on the battlefield and virtually stationary in combat. The 75mm Type 38 (Improved) fired a 10.6kg (23.4lb) shell a maximum range of 11,960m (13,080yds).

When war broke out with the United States, the Japanese forces in the Pacific found themselves routinely outgunned. Indeed, the sustained firepower of American artillery shattered the Banzai charges of Japanese infantry on more than one occasion and inflicted heavy casualties upon the attackers. In strengthening their island defences, the Japanese did later avail themselves of heavy guns captured during their early successes. Weapons were moved from fortresses in Singapore,

During the invasion of China, Japanese soldiers load a light artillery piece. Generally, the Japanese advanced so quickly that little reliance was placed on heavy weapons.

HEAD TO HEAD: *105mm Howitzer M2A1 VERSUS*

With more than 8000 produced during the course of World War II, the 105mm howitzer M2A1 was the primary field weapon of the American artillery units. The design was completed in 1939, and production began the following year.

105mm Howitzer M2A1

Calibre: 105mm (3in)
Weight in action: 2030kg (4475lb)
Gun length: 22 calibre: 2.31m (91in)
Elevation: -5° to +66°
Traverse: 46°
Shell type & weight: HE; 15kg (33lb)
Muzzle velocity: 472m/sec (1548ft/sec)
Maximum range: 11,200m (12,248yds)

STRENGTHS

- Excellent availability
- Durability and longevity
- Lasting design influence

WEAKNESSES

- Heavy weight
- Truck towed
- Limited mobility

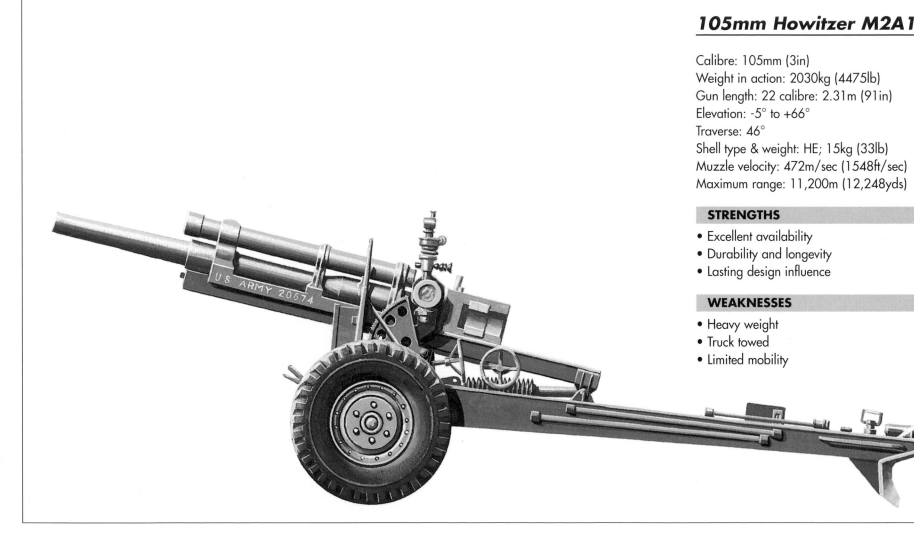

Type 35/75mm Gun

A prime example of lagging Japanese artillery development, the Type 35/75 Gun was built in Japan under licence from Krupp and based on a design that dated back to 1905. The weapon was greatly inferior to its Allied counterparts.

Type 35/75mm Gun

Calibre: 75mm (2.95in)
Weight in action: 1007kg (2.95in)
Gun length: 31 calibre: 2.3m (91in)
Elevation: -8° to +43°
Traverse: 50°
Shell type & weight: HE; 5.7kg (13lb)
Muzzle velocity: 520m/sec (1700ft/sec)
Maximum range: 10,700m (11,700yds)

STRENGTHS

• Simple design
• Available in large numbers
• Reliable

WEAKNESSES

• Antiquated performance
• Animal transport only
• Lack of firepower

for example, to strengthen the defences of Tarawa in the Gilbert Islands.

The dogs of war

When Hitler's Blitzkrieg rolled across the Polish frontier, the German Army was the most modern and best equipped in the entire world. Artillery worked in close coordination with infantry and air power to rapidly seize territory and deprive the Polish Army of its communications capability. Images of devastation flashed across the world, and once again the nations of Europe were plunged into armed conflict.

Recollections of the stagnation of the Western Front in the previous war and the horror of the trenches placed a new priority on the tactical employment of artillery for the war planners of some nations. Mobile firepower, combined with speed and accuracy of fire, could be deciding factors in the second global conflict of the twentieth century, they reasoned. For the military establishments of some nations, therefore, a growing sense of urgency was marked by substantial progress and modernization. For others, either refusing to come to terms with the changing face of war or failing to grasp its significance, the bitter shortcomings of the 1920s and 30s were subsequently laid bare.

Right: Advancing into the Soviet Union, German artillerymen ride aboard the horse-drawn carriage of their heavy field gun.

VICTORY IN EUROPE AND THE PACIFIC

On 10 June 1941, President Franklin D. Roosevelt addressed the United States Congress. 'With our national resources, our productive capacity, and the genius of our people for mass production,' he declared to the assembly, 'we will ... outstrip the Axis powers in munitions of war.'

Indeed, the industrial capacity of the United States and the Soviet Union provided the materiel of war necessary to bring their enemies to their knees. By the end of World War II, American factories had produced 372,431 artillery pieces, 42 million tonnes of artillery projectiles and 102,351 self-propelled guns and tanks. Additionally, during the war years, nearly 4500 field artillery pieces were provided to Allied countries via the Lend-Lease programme. At the Teheran Conference in 1943, the monumental mobilization of American industry was acknowledged by Soviet Premier Joseph Stalin, who raised his glass in tribute, saying, 'To American production, without which this war would have been lost.'

Left: On a Pacific island, American artillerymen prepare for a renewal of battle. Concentrated American artillery fire decimated Japanese banzai charges.

Uncharacteristically, Stalin did not drink to his own nation's incredible effort to supply its armed forces. In *Why the Allies Won*, Richard Overy relates the stirring story of the relocation of aircraft designer Alexander Yakovlev and his bureau, following the dismantling of the production facilities, which were in danger of being overrun by the advancing Nazis:

'What Yakovlev witnessed was a small part of a vast exodus. Between July and December 1941 1523 enterprises, the great bulk of them iron, steel and engineering plants, were moved to the Urals, to the Volga region, to Kazakhstan in central Asia, and to eastern Siberia. One and a half million wagon loads were carried eastwards on the Soviet rail network. An estimated 16 million Soviet citizens escaped the German net, many of them factory workers, engineers, plant managers, all needed to keep the uprooted industries going.'

In 1943, these relocated production facilities manufactured 48,000 heavy artillery pieces and 24,000 tanks. By 1945, when the Red Army stood poised to strike at Berlin, Marshal Georgi Zhukov was able to muster 41,600 artillery pieces to bombard the German capital.

Compared to Soviet production in 1943, the Germans manufactured 27,000 heavy guns and 17,000 tanks. Under the duress of Allied bombing and the strain of fighting a war on two fronts, German industry actually did achieve impressive production figures. Meanwhile, the industrial capacity of Japan was being systematically strangled by American submarines, which sent millions of tonnes of raw materials to the bottom of the Pacific Ocean with torpedoed merchant shipping.

Although the weight of Allied arms and manpower eventually began to tip the balance of power and shape the outcome of World War II,

An American Boeing B-17 Flying Fortress bomber streaks away from its target, which is seen billowing smoke in the distance.

great land battles at El Alamein, Stalingrad and Guadalcanal were fought with grit and determination. Countless engagements, large and small, took place before the guns fell silent.

To command the skies

During World War II, the combat aircraft came of age as both a strategic and tactical weapon. Gradually, Allied air forces gained supremacy in the skies over Europe and Japan. Formations of heavy and medium bombers dropped destructive payloads on the homeland of the Third Reich, while fighter-bombers harassed the movement of troops and armoured columns and disrupted the flow of supplies. Defensively, the anti-aircraft weapon provided both Allied and Axis forces with some measure of response to the threat of attack from the air. Both heavy and light anti-aircraft weapons were deployed in large numbers and a variety of calibres.

One major wartime advance that improved the effectiveness of anti-aircraft weapons was the introduction of the proximity fuse. Developed initially in secret by a joint team of American and British scientists for use against enemy warplanes, the proximity fuse contained a small radio transmitter. This emitted a signal intended to be reflected by a nearby solid object, which was then picked up by the fuse. The exchange of the signal triggered the detonation of the shell. Therefore, a direct hit would not be required to inflict serious damage. Later, the proximity fuse was adapted for use with field artillery as well.

The US Army implemented the concept of the fire direction centre in the 1930s. The fire direction centre involved a forward observer using a radio, thus eliminating the cumbersome field telephone and its accompanying wire, or reporting from a light observation aircraft. The observer was able to locate distant targets out of sight of the gun crews themselves. The fire direction centre also served as a central point for the gathering of data to plot the coordinates and place fire on to a target.

Theoretically, the concept could bring to bear all the artillery firepower of a division or corps against a single target within 10 minutes of the initial call from the forward observer.

A curtain of steel

German anti-aircraft units were typically under the command of the air force, or Luftwaffe. Highly trained crews were capable of producing a tremendous volume of fire while protecting German cities, military installations or troop concentrations. Anti-aircraft weapons were often mobile, although some were fixed in stationary ground positions or flak towers, which offered more favourable firing positions surrounding high-value targets.

Many pilots and crewmen of the Eighth US Air Force and the British Royal Air Force braved the gauntlet of steel. Veterans recall that the sound of shell fragments striking was similar to that of a handful of stones being thrown against their plane's aluminium skin. While German fighter planes were a fearsome threat, the likelihood of being shot down by flak was generally much greater. During 1945, for example, the number of US fighter aircraft lost to ground fire was greater than that lost to enemy fighters by a significant margin of four to one.

Leroy Faulkner, a waist gunner aboard a Boeing B-17 Flying Fortress heavy bomber of the Eighth Air Force, remembered:

'On a clear day when contrails could be seen, somebody was going to be killed. The Germans would get our altitude and speed along with our range, and then it got nasty. The flak would come in full force from the 88mm guns. They would literally shoot holes through us. We always wore flak helmets and vests.'

Lieutenant Colonel James R. Maris piloted a Consolidated B-24 Liberator of the 392nd Bomb Group, 578th Bomb Squadron, during a harrowing mission over Hamburg in August, 1944. He recalled:

'"Engineer to pilot, engineer to pilot: Our number one engine has been blown off the wing.

Responding to the alarm of a coming Allied air raid, German anti-aircraft gunners spring to their 88mm flak guns adjacent to the rail yard they are defending.

Number three is stripped of its cowl and supercharger. There's a three-foot wide hole in the left wing between engines one and two. The bomb bay doors are crushed in. And we've got a bomb hung up on the shackles in the bomb bay." This was the frantic report from my flight engineer, Milford 'Fitz' Fitzgerald, who had been asked to assess the damage to our B-24, the *Bad Penny*. She had been badly damaged as we passed through heavy flak … on our 23rd mission.

'We entered the flak storm over our target and were immediately tossed by the severe turbulence created by the exploding 88-mm flak. The *Bad Penny* was bathed in brilliant flashes of light and peppered with exploding shells. She rocked and shuttered [sic] with the jarring impact of every burst. The biggest jolt came when the number one engine was blown off. We rolled hard to the right and it was all that my copilot and I could do to right our B-24. Not long after, a second blast stripped the cowling and supercharger off engine

number three on our right wing. An engine oil fire created an expanding plume of white smoke that trailed our aircraft.'

Defending the Reich

While the 88mm series of anti-aircraft guns was well known for its prowess, both in its original role and as an anti-tank weapon, the Germans were by far the leaders in the variety of heavy and light anti-aircraft guns deployed during World War II. The consensus among German armaments experts was

that an anti-aircraft weapon of heavier calibre than the 88mm was needed, and by the mid-1930s, the Rheinmetall 105mm Flak 38 had been selected for production following a competition with Krupp. The Flak 38 was quickly superseded by the Flak 39, which incorporated an improved fire-control data system.

The major shortcoming of both models lay in their considerable weight. The Flak 39 was a hefty 10,240kg (22,575lb) when firing. Coupled with the fact that the weapon's performance was not appreciably better than that of the 88, the weight problem relegated the 105mm models to primary use in fixed fortifications. More than 100 were also mounted on specially designed railway trucks. The Flak 39 was manufactured with a sectional barrel, which complicated the process and slowed production. Firing a 15.1kg (33.3lb) projectile to a ceiling of 12,800m (41,995ft), the 105mm Flak 38 and 39 were nevertheless disappointing in their overall performance and were never used as field guns on a large scale. Another version of the weapon, the Flak 40, was discarded during the research phase, and more than 1800 of the Flak 38 and 39 were still in operation at the end of the war.

The prototype of the Rheinmetall 128mm Flak 40 was introduced in 1940, and regular production began two years later. Generally known as the Gerät 40, it, too, was intended originally for use in the field. However, its static firing weight of 12,300kg (28,660lb) was unsuitable for rapid movement, and this weapon was also primarily used to defend major cities. At times, it was installed in flak towers or placed on railway cars. At the end of the war, a total of just 570 Gerät 40s were in service, firing a 26kg (57.3lb) shell up to 14,800m (48,555ft).

105MM FLAK 38

Calibre: 105mm (4.13in)
Weight in action: 10,224kg (10.06 tons)
Gun length: 6.648 (21ft 10in)
Elevation: -3° to +85°

Traverse: 360°
Shell type & weight: HE; 14.8kg (32.6lb)
Muzzle velocity: 881m/sec (2891ft/sec)
Effective ceiling: 9450m (31,000ft)

Fixed atop a self-propelled halftrack chassis, the quadruple-mounted 20mm Flakvierling 38 was one of the most effective German light anti-aircraft weapons.

Light anti-aircraft guns were produced in large numbers by the Germans. By the end of 1940, the 20mm Flak 38 had entered service as a replacement for the slower-firing 20mm Flak 30, which had also proven to be an overly complicated weapon to operate due to its size. In the Flak 38, Mauser had produced a weapon that had a rate of fire up to 480 rounds per minute and which was better capable of coping with the greater speeds of high-performance aircraft being introduced by the Allies. Also in 1940, the Flakvierling 38 was introduced, its four barrels considered the most expedient method of compensating for the lack of knockdown capability in a single-barrelled weapon. The Flak 38 fired a 0.125kg (4oz) shell to a maximum ceiling of 2200m (7218ft), while the Flakvierling 38 increased the rate of fire to a substantial 1800 rounds per minute. Pilots of low-flying Allied aircraft, particularly fighter-bomber pilots on strafing runs, came to respect both weapons. The German Navy deployed its own version, and a number were mounted on flak trains. The weapons were continually in short supply.

The somewhat heavier 37mm Flak 18 family was developed by Rheinmetall in the early 1930s in Switzerland to circumvent the terms of the Versailles Treaty. The Flak 18 entered service in 1935 but was plagued with mechanical failures and a heavy carriage, which made movement and traverse both slow. When production of the Flak 18 was discontinued in 1936, a similarly configured weapon, the Flak 36, was introduced. Although the two weapons appeared identical, the Flak 36 had been mechanically improved and could be towed on a single axle for better mobility. The following year, the 37mm Flak 37 was introduced with improved sighting equipment.

From 1940 onwards, the Flak 18 family of guns became the primary light air defence weapons of the German armed forces, deployed in batteries of up to 12 guns. The German Navy used the weapon on submarines, while others were placed in flak trains or towers. By the autumn of 1944, more than 4200 of the guns were in service. At times, it was also pressed into an anti-tank role.

The 37mm Flak 43 and Flakzwilling 43 arrived in combat areas in early 1944, having been produced in a quarter of the time required for previous models, due to the adoption at Rheinmetall of manufacturing techniques similar to those applied to small arms. An improved rate of fire was an additional benefit with the Flak 43. However, the rush to produce them meant that it was necessary to stay with existing 37mm ammunition, which lacked the knockdown capabilities desired. Often an Allied plane would sustain damage but stay in the air. Therefore, the Flakzwilling 43 continued the concept of a multi-

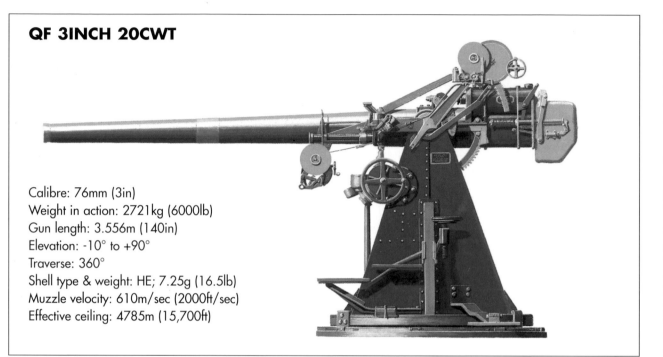

QF 3INCH 20CWT

Calibre: 76mm (3in)
Weight in action: 2721kg (6000lb)
Gun length: 3.556m (140in)
Elevation: -10° to +90°
Traverse: 360°
Shell type & weight: HE; 7.25g (16.5lb)
Muzzle velocity: 610m/sec (2000ft/sec)
Effective ceiling: 4785m (15,700ft)

barrelled weapon. Its two cannon, mounted one below the other, increased the likelihood of a shootdown. Production constraints were considerable, and by the end of the war just over 1000 Flak 43 and a scant 280 Flakzwilling 43 weapons were in service. The guns fired 0.64kg (1.41lb) projectiles up to a 4800m (15,748ft) ceiling.

Allied aircraft flew in a zone from 1494m (4900ft) to 3048m (10,000ft), which was either below or above the effective range of heavy and light weapons. In order to attack effectively, the 50mm Flak 41 was introduced in mobile and static variants. Its crew of seven men could fire 180 rounds per minute, but the performance of the weapon was hampered by under-strength

ammunition, which produced excessive muzzle blast. Sixty of the weapons, which fired a 2.2kg (4.85lb) shell up to slightly more than 3048m (10,000 feet), were placed in service, but by the end of the war only 24 remained operational.

The best gun of the war

If there was such a thing as a universally acclaimed anti-aircraft weapon during World War II, it was a gun that had been developed in a neutral country. The Swedish 40mm Bofors traced its origins to a request from the Swedish Navy, and it was being manufactured under licence throughout Europe by the mid-1930s. The weapon provided a high rate of fire, 120 rounds per minute, and was manufactured in a single-barrel mobile mounting for land forces

and a dual configuration for naval applications. The weapon's high muzzle velocity, 854m (2802ft) per second, contributed to its success as an anti-aircraft gun. When representatives of Bofors visited foreign countries in the 1930s to demonstrate the weapon, their usual comment was, 'One shot, and the contract is ours.'

The 40mm Bofors could be fired virtually as an automatic weapon because of its continuous feed capability, the spent case being ejected and a new round thrust into the breech until ammunition had been exhausted from the clip. The Germans manufactured the 40mm Bofors in Norway and designated it the 40mm Flak 28 (Bofors), while the United States fabricated it as the 40mm Gun M1. Large numbers of them were also produced by the British Commonwealth. The weapon was in service in every theatre of World War II, firing a 0.9kg (1.9lb) shell to a maximum ceiling of 7200m (23,622ft). A 20mm variant was also developed during the 1930s.

Heavier Bofors weapons, in 75mm and 80mm calibres, were also produced in numbers from the 1930s onwards. These weapons closely resembled the famed 88mm gun produced by Krupp after its joint venture with Bofors during the 1930s to produce the 75mm weapon. Bofors, however, insisted that the 75mm and 80mm weapons were wholly Swedish designs. Both weapons were exported to numerous countries, and the 80mm saw wide use on the Eastern Front with the Hungarian Army. The 80mm version fired an 8kg (17.6lb) shell up to 10,000m (32,810ft).

Weapons from an earlier war

The grandfather of British anti-aircraft weapons was the venerable Ordnance, QF, 3 in 20 cwt,

5CM FLAK 41

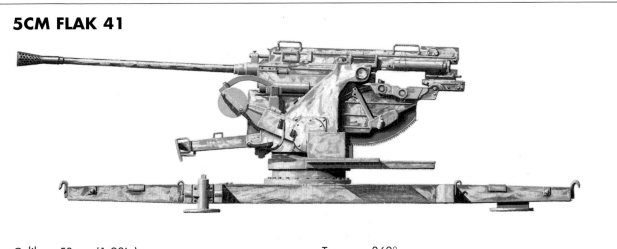

Calibre: 50mm (1.98in)
Weight in action: 3100kg (6838lb)
Gun length: 4.69m (185in)
Elevation: -10° to +90°

Traverse: 360°
Shell type & weight: HE; 22.25kg (4.85lb)
Muzzle velocity: 480m/sec (3756ft/sec)
Effective ceiling: 5600m (18,375ft)

Some of the 3in guns found extended life in Home Guard units and coastal defences, few of them remaining active by the end of the war. The 3in gun fired a 7.25kg (16lb) shell up to a ceiling of 7163m (23,500ft).

Within months of the end of World War I, a report had suggested that a heavier anti-aircraft weapon should be produced in Britain to replace the 3in gun. However, severe financial constraints postponed the development of such a weapon until the mid-1930s when Vickers produced the first Ordnance, QF, 3.7in Mark I, which was capable of firing a 12.9kg (28.4lb) projectile to a ceiling of 9754m (32,000ft). The large weapon weighed 9276kg (20,451lb), and the army's requirement that it be mobile could only partially be met due to issues surrounding the carriage, which was improved in the subsequent Mark II and Mark III versions as the war progressed. Although it was slow

which had originally been placed in service with the army as early as 1914. The 3in weapon was, by 1939, widely in use as a static and mobile gun, and it was deployed to the continent with the British Expeditionary Force in 1939. By the beginning of World War II, the 3in gun existed in numerous configurations, including a variety of breechblocks and carriages. While troops in the field preferred the lighter weapon over its proposed replacement, the 3.7in cannon, most of the guns were abandoned during the evacuation at Dunkirk in the summer of 1940 and captured by the Germans who renamed them the 75mm Flak Vickers(e).

A US Marine gun crew mans its Swedish-made 40mm Bofors anti-aircraft gun on the island of Bougainville in the Pacific Ocean.

HEAD TO HEAD: *40mm Bofors L/60* VERSUS

Originally designed in 1928 at the request of the Swedish Navy, the 40mm Bofors gun is a legendary weapon. Manufactured in several countries under licensing agreements, the Bofors was used by both Allied and Axis forces.

40mm Bofors L/60

Calibre: 40mm (1.57in)
Weight in action: 1981kg (4568lb)
Gun length: 3m (118in)
Elevation: -5° to +90°
Traverse: 360°
Shell type & weight: HE; 907g (2lb)
Muzzle velocity: 823m/sec (2700ft/sec)
Effective ceiling: 7200m (23,622ft)

STRENGTHS

• High muzzle velocity
• Large projectile
• Light weight

WEAKNESSES

• Machine towed
• Relatively small calibre
• Widespread use

Flak 43

The German 37mm Flak 43 and the Flakzwilling 43 (shown) were developed by Rheinmetall-Borsig as the standard German anti-aircraft weapon for use against low-flying Allied aircraft. Its development was plagued by political infighting. The Flakzwilling was an upgunned version of the Flak 43, adding a second barrel.

Flak 43

Calibre: 37mm (1.45in)
Weight in action: 1248kg (2752lb)
Gun length: 3.3m (130in)
Elevation: -7.5° to +90°
Traverse: 360°
Shell type & weight: HE; 635g (22.4oz)
Muzzle velocity: 820m/sec (2690ft/sec)
Effective ceiling: 4200m (13,780ft)

STRENGTHS

- Rapid production
- High ceiling
- High rate of fire

WEAKNESSES

- Low firepower
- Unwieldy design
- Limited availability

to gain the appreciation of gun crews, the 3.7in cannon ultimately proved to be an outstanding performer. By 1941, it was the primary anti-aircraft weapon in use by the British Army. The Germans respected the weapon highly, used those they captured in coastal defences, renamed it the 94mm Flak Vickers M.39(e), and even manufactured ammunition for it.

Originally intended as a naval weapon only, the Ordnance, QF, 4.5in, AA Mk II, was adopted by the army in the late 1930s, but only with the cooperation of the Admiralty. The naval hierarchy had stipulated that the weapon be deployed around high-value targets such as dockyards and anchorages. The heavy weapon, weighing more than 16,783kg (37,000lb) in static emplacement, was mobile only on a limited basis by means of a specially configured transportation system. Firing a 24.7kg (54.5lb) shell up to 12,984m (42,600ft), the 4.5in gun was sometimes moved to coastal defensive positions late in the war; however, by the 1950s it was being phased out of service.

The 20mm Polsten light anti-aircraft gun was produced only in the British Commonwealth during the war years, but it was originally of Polish design. When Germany overran Poland in 1939, the designers of the weapon made their way to Britain and continued to work on the gun, which, in truth, was a refinement of the Swiss 20mm Oerlikon. The Polsten was considerably less expensive to build than the Oerlikon, with the number of components reduced by more than half. The Polsten reached the front lines in March 1944, and could be fitted to any carriage that could mount the Oerlikon. It also found use as a tank-mounted weapon and aboard aircraft. The gun's 0.125kg (4oz) shell had an effective ceiling of 2021m (6630ft).

A full battery of 12 British Q.F. 3.7in anti-aircraft guns fires skyward at an extremely high trajectory. The occasion was a celebration of victory in Europe, May 1945.

The 20mm Oerlikon was initially designed by a German, Reinhold Becker, during World War I. However, when the war ended, Becker transferred the weapon's manufacture to Switzerland. Several types were produced, and the gun was also manufactured under licence in numerous countries, including Great Britain and France. It fired an 0.125kg (4oz) shell a maximum ceiling of 1097m (3600ft). Ammunition was normally supplied to the gas-operated gun from 20- or 60-round boxes. The Oerlikon was widely used aboard aircraft and warships of virtually every warring nation. The Germans designated the Oerlikon as the 20mm Flak 28 or 29, while the Italians renamed it the Cannone-Mitraglia da 20 Oerlikon, and the Americans the 20mm Automatic Gun Mk IV.

One of the first Allied land weapons to fire the proximity fuse in combat was the American 90mm Gun M1, which had been proposed as a replacement for the obsolescent 3in weapon in 1938 with approval of the prototype coming in 1940. Both the 90mm gun and its carriage were difficult to produce due to very specific machining tolerances and towing challenges. An improved version, the M1A1, appeared in early 1941 and incorporated what proved to be a troublesome spring rammer. More revisions quickly followed, and eventually nearly 8000 90mm guns were placed in service. The first use of the proximity fuse occurred during the Battle of the Bulge in December 1944. The 90mm weapon fired a 10.6kg (23.4lb) shell to a maximum ceiling of 12,040m (39,500ft).

The older 3in M3 anti-aircraft gun had gone into production near the end of World War I but was recognized as outdated by the mid-1930s. Several of these were used in action against the Japanese prior to the fall of the Philippines in the spring of 1942. Later, it was primarily used for training purposes, firing a 5.8kg (12.8lb) shell up to a ceiling of 9510m (31,200ft).

The 37mm M1 and the Maxson Mount were two prominent American-made light anti-aircraft weapons of World War II. The Maxson was a mobile configuration of four .50-calibre Browning machine guns and is included here due to its primary role against low-flying Axis aircraft and the

concentrated firepower produced by four heavy machine guns spewing 2300 rounds per minute. The highly mobile Maxson was placed on halftracks or trailers and was electrically operated. Its ceiling was 1000m (3280ft), and at times the

weapon was pressed into service against enemy infantry formations with devastating effect.

The 37mm M1 was first proposed in 1921 but languished for a number of years before research was resumed. By 1940, the Colt Company had

Utilized primarily as a naval weapon the Swiss-designed 20mm Oerlikon also saw service with land forces. Here, sailors aim their weapons skywards aboard a warship.

begun production of the design originally conceived by John M. Browning. Improvements resulted in the M1A2 variant, but evaluations of the gun's performance quickly indicated that the 40mm Bofors was superior. Over time, production was fully changed to the Bofors, and numbers of the 37mm M1A2 were transferred to the Soviet Union via Lend Lease. Another configuration of the 37mm combined two 50mm calibre machine guns mounted on either side of the cannon and designated the Combination Mount M54. The 37mm weapon fired a 0.6kg (1.3lb) projectile up to 5670m (18,600ft) at a rate of 120 rounds per minute.

The primary Soviet anti-aircraft gun of World War II was the 85mm cannon, which was an upgrade of the tried and true 76.2mm design of World War I. Introduced in 1939 and replaced by a more powerful variant in 1944, the 85mm cannon fired a 9.2kg (20.2lb) shell up to 10,500m

The Lend-Lease Lifeline

In March 1941, eight months prior to US entry into World War II, President Franklin D. Roosevelt signed House Resolution 1776, the Lend-Lease bill that would eventually allow for the transfer of millions of dollars in war materiel to the beleaguered nations of Great Britain, the Soviet Union, China, France and others battling the Axis powers. Before the Lend-Lease programme was terminated in 1945, more than $50 billion worth of supplies and equipment, the equivalent of $700 billion in modern terms, had been exported from the United States to its allies. Among the items shipped abroad were more than 8,000 pieces of artillery, 15,000 aircraft, four million tyres and 15 million boots.

(34,450ft). It was also used as an anti-tank weapon, and thousands of the 85mm gun were produced and remained in service around the world for decades.

The Japanese Type 88 75mm gun dated from the late 1920s and was unsatisfactory by the time large formations of American bombers were flying high above the home islands to deliver their payloads. Its

An American 90mm M1 anti-aircraft gun is shown mounted on the early M1 carriage. Other emplacements can be seen dug in along a beach to provide defensive fire.

maximum ceiling of 7250m (23,785ft) was too low for effective fire, and its 6.5kg (14.3lb) shell lacked the necessary power. Hampered by a lack of research and development, as well as limited

Dug into a low profile position in the Solomon Islands in the Pacific, this formidable Maxson mount is typical of those deployed with American forces.

production capability, the Japanese were forced to press a variety of naval cannon into service as the war progressed. The Type 98 20mm machine cannon was effective against low-flying Allied aircraft, easily disassembled for movement by pack animals, and widely used during the war, firing at 120 rounds per minutes to a ceiling of 3650m (11,975ft). A 25mm weapon was also placed in service.

Other anti-aircraft weapons of note included the Italian 75/46 Model 34, a 75mm weapon and the 90mm Cannon 90/53. A pair of light Italian weapons were the highly regarded 20mm Breda and the 20mm Scotti. The French 75mm gun was used by various forces in an anti-aircraft role throughout the war, and the light 25mm Hotchkiss and 37mm Schneider guns saw service as well.

The battling bastards

Although Japanese military doctrine has been maligned for its lack of attention to artillery during World War II, the country's armed forces did deal effectively with fortified positions at Hong Kong and Singapore early in the conflict. During the Philippine campaign of 1941–42, the Japanese were reported to have brought in a variety of artillery, ranging from comparatively light 75mm mountain guns to mammoth 320mm spigot mortars. Working cooperatively with large numbers of troops on the ground and leveraging complete air supremacy to bomb and strafe at will, they pounded the fortifications of Bataan and Corregidor into submission.

Records indicate that Japanese artillery fired no less than 63,000 rounds against the fortified American positions on Bataan and Corregidor, which included more than 60 heavy weapons up to 14in calibre. Fort Drum, occupying a small island in Manila Bay, was referred to as 'the concrete battleship' with both 6in and 14in weapons. While a number of the big American guns on Bataan and Corregidor dated from the late nineteenth century and included the archaic disappearing mounts, they remained a threat. However, the sheer weight of numbers spelled defeat for the defenders, who came to be known as the 'Battling Bastards of Bataan', and resulted in the largest capitulation of American forces in history.

Later in the war, the Japanese did remove heavy weapons from conquered territory and place them in defensive positions on strategically important islands in the Pacific. At Tarawa, the US destroyers Ringgold and Dashiell were seen sweeping into the shallow lagoon of the atoll to duel Japanese shore batteries with their 5in naval rifles. At Peleliu, Iwo Jima and Okinawa, the Japanese placed heavy guns in the shelter of caves and bunkers, sometimes revealing themselves only long enough to register a target and fire before retracting out of sight. Often, naval gunfire and air attacks were unsuccessful against such positions, which had to be neutralized by individual soldiers with satchel charges and flamethrowers. Iwo Jima veteran Charles W. Tatum, of the US Marine Corps, recalls:

'What I thought was another mortar shell fell on the same spot as before ... the blast's shock wave

On the island of Tarawa in the Gilberts, American Marines inspect captured 8in cannon that the Japanese had seized at Singapore and transported to Tarawa.

whipped up black dirt which pushed its way into my eyes and forced sand into my mouth – gagging me. Only when the sand and dust cleared, could I see [Marine Gunnery Sergeant John] Basilone was pointing at the aperture of a reinforced concrete bunker or blockhouse. The structure probably housed a 75mm or larger cannon whose field of fire was directed down the beach to our right. It was a big bastard with incredible killing power.'

On several occasions during the arduous trek of US forces across the Pacific, massed Banzai charges of Japanese infantry were broken up by concentrated artillery fire, decimating the attacking

forces. At Saipan in the Mariana Islands, a tide of Japanese soldiers swept into the American lines.

Captain Edmund G. Love recalled:

'With the disruption of communications and the movement of battalions under the trees of the village, they were largely out of sight of the main division observation posts (OP's). But the men manning these OP's could see the plain quite easily and watch the Japanese moving around openly in the whole area. Artillery liaison planes had been flying overhead all morning, picking out targets. Cannon Company, 165th Infantry, led by Capt. Robert B. Marshall, had perched itself upon the shelf of ground under a hilltop, and the gunners threw shells into the swarms of Japanese below.'

Festung Europa

As the prospects faded for the invasion of Great Britain, codenamed Operation Sea Lion, workers of the Todt Organization and forced labourers began constructing the fortifications of the Atlantic Wall, described by Hitler as 'Festung Europa', or Fortress Europe. Augmenting existing French defences, the Germans intended to construct such a formidable obstacle to Allied landings that an invasion of the continent would be deemed too costly.

To that end, a major construction initiative was commenced in 1942. During the months of January and July alone, more than 700,000 cubic metres (915,565 cubic yards) of concrete were poured. Bunkers, blockhouses and reinforced gun emplacements sheltered a variety of artillery, and on the eve of Operation Overlord, the Allied invasion of occupied France on 6 June 1944, the Germans had installed along the French coastline between Dunkirk and Le Havre no fewer than 120 heavy

A German soldier peers from a firing port past the barrel of a light artillery weapon. Hitler's Atlantic Wall, built along the coast of France, was studded by such positions.

batteries of guns, ranging from 100mm to more than 210mm. Along the Normandy coast to the west of the River Seine were up to 50 guns of similar size. Many of these were housed in strong points virtually impervious to air attack and prolonged naval bombardment.

So concerned were Allied military leaders that enemy artillery might wreak havoc on the D-Day invasion beaches that high-risk operations were planned to eliminate these perceived threats. Two

such missions included the predawn capture by British airborne troops of the Merville Battery and the assault of 225 US Rangers, scaling the sheer, 30m (100ft) cliffs at Pointe du Hoc to silence another potentially lethal position. The Merville Battery, situated in a commanding position above Sword Beach, housed powerful 100mm guns in a concrete casemate fully 2m (6.6ft) thick. A garrison of 160 soldiers defended the battery, which was ringed on the landward side by barbed wire, trenches, machine gun nests and three 20mm anti-aircraft guns. The job of taking the position fell to the British 9th Battalion of the Parachute Regiment, commanded by Lieutenant Colonel

Terence Otway. Otway began the mission with more than 650 troops, but navigational errors and weather conditions caused the gliders to be dispersed as far as 16km (10 miles) from the landing zone. Otway could scrape together only 150 men to assault the strong German position, and had virtually no heavy equipment. Overcoming fierce resistance, the glider-borne troops captured the battery. The price, however, was expensive, with 85 of the men engaged killed or wounded in the action.

Otway later remarked:

'It's been estimated that we saved hundreds if not thousands of lives. They [the 9th Battalion] were magnificent, all of them. What's that they say about the British when they've got their backs to the wall?'

Using grappling hooks and ladders borrowed from London firefighters, three companies of the US 2nd Ranger Battalion fought their way to the top of Pointe du Hoc, where a battery of six 155mm guns supposedly was housed in a concrete reinforced bunker. The position had been pounded by bombs and naval gunfire, but there was no way of knowing for certain that the German 155s had been put out of action. If operational, they could enfilade both Omaha and Utah beaches with heavy fire.

Once the Rangers had driven the Germans back and scaled the cliffs, however, the guns were nowhere to be found. One Ranger noticed tracks from wheeled vehicles in the mud, and the guns were later located under a canopy of trees in an orchard a considerable distance from the invasion beaches. The Rangers smashed the sights with the butts of their rifles and dropped thermite grenades down the barrels to permanently disable the

Rangers at the Pointe

Approximately 200 men of the US 2nd Ranger Battalion endure rugged training in preparation for the D-Day assault against the cliffs of Pointe du Hoc between the Utah and Omaha invasion beaches. The potential presence of heavy German artillery necessitated a dangerous mission to destroy the guns.

weapons and end the threat. When the fighting around Pointe du Hoc was over, only 90 Rangers remained uninjured.

Battling the armoured beast

Although practical research and development on anti-tank weapons did not flourish until the 1930s, there was no denying that the tank was a formidable weapon on the battlefield. The German Blitzkrieg of 1940 demonstrated the tremendous capability of armoured formations to create and exploit breakthroughs in enemy lines. Most anti-tank weapons of the early war period were of light 20mm and 40mm calibres. However, advances in firepower and mobility among tanks necessitated that the size and number of these weapons be increased to deal with the threat effectively.

Improvements in ammunition increased the lethality of anti-tank guns throughout World War II. Initially, the solid projectiles in use were not powerful enough to penetrate thick armour, and the most expedient remedy was to increase muzzle velocity. However, this practice caused considerable wear on gun barrels and proved impractical. Both British and German weapons researchers developed shells, which used projectiles of dense tungsten and tungsten carbide, travelling at a high muzzle velocity. The German shell included a tungsten core surrounded by a malleable metal. Tapered bore guns specially designed to fire these shells were also developed. The British tungsten projectile was delivered by means of a discarding sabot shell, consisting of a tungsten core surrounded by a lighter cover, which fell away after the projectile cleared the barrel.

Another innovation was the shaped or hollow charge. These shells were distinguishable by their conical or flat-headed and bulbous appearances. Upon impact, these charges either penetrated armour by forming a molten metal jet that bored through armour or caused the failure of the steel by generating a substantial shock wave.

Kursk – the great clash of armour

Historians describe the Battle of Kursk in the summer of 1943 as the greatest clash of opposing armoured forces in history. Thousands of Soviet medium T-34 and heavy KV-1 tanks battled the German medium Panther and heavy Tiger tanks of the Waffen-SS at Prokhorovka, which came to be called the 'Gulley of Death'. The dual purpose of many artillery pieces deployed by the opposing forces was evidenced on numerous occasions.

One German soldier recalled:

'On the second day of the operation, the high ground ahead was captured under the protection of Flak artillery fire. All counterthrusts were repelled. On the morning of the third day the enemy attempted to recover lost ground and counterattacked with two heavy tank brigades and motorized infantry units. The tanks overran the battle line of the German infantry and penetrated deeply, but the motorized infantry which followed was repelled. The enemy tank breakthrough hit the corps centre, behind which, however, several Flak assault detachments and numerous medium anti-tank guns were sited in a mutually supporting formation.

'The enemy ran into this dense network of anti-tank defences as well as a flank attack by 32 assault

The largest tank battle ever fought, the Kursk in summer 1943 also included heavy barrages by German and Soviet Red Army artillery.

guns and was completely annihilated. The last enemy tank which had penetrated to a divisional command post was surprised by an assault detachment carrying gasoline cans and was set on fire. Sixty-four enemy tanks had begun the

counterattack and two hours later 64 black columns of smoke gave proof of their destruction … the improvised commitment of the anti-aircraft division contributed decisively to this defensive success and the formation of Flak assault

detachments proved highly effective in the destruction of Russian armour.'

As early as 1925, Rheinmetall researchers had begun working on an anti-tank weapon, and by the time the Nazis came to power eight years later, some anti-tank guns had been placed in service with the Reichswehr, the pre-Nazi German Army. By 1936, the Rheinmetall design had been upgraded with pneumatic wheels for its carriage and formally designated the 37mm Pak 35/36. The weapon saw service during the Spanish Civil War, but it became apparent by 1940 that it was too weak to penetrate the thicker armour of the more modern British and French tanks. Its performance against Soviet tanks during Operation Barbarossa was also dismal. Thereafter, it was relegated to service in training units. A number of the weapons were exported, and it was notably produced directly by Japan and designated the Type 97. The Pak 35/36 fired a 0.6kg (1.3lb) projectile a maximum range of 6700m (7655yds).

Responding to the obvious need of a more powerful anti-tank gun, the Germans fielded the 50mm Pak 38 in the summer of 1940, but the weapon did not arrive in numbers great enough to be decisive during the invasion of the Soviet Union. However, it did perform well against the thick armour of the Red Army T-34 tank, particularly when firing new AP40 ammunition with a tungsten core. The Pak 38 provided cover for its crew with a curved shield and was adapted for automatic feed, resulting in a weapon that was popular with the troops and produced throughout the war. Interestingly, the Messerschmitt Me-262 jet fighter aircraft was armed with a variant. The AP40 ammunition of the Pak 38 was capable of

penetrating 10cm (3.98in) of armour at a distance of 750m (820yds).

On the heels of the Pak 38 came the upgunned 75mm Pak 40, which could also fire the AP40 round, penetrating 9.8cm (3.86in) of armour at 2003m (2190yds). The 75mm Pak 40 was significantly heavier than that Pak 38 due to shortages of lighter steel alloys, but the weapon remained in production until the end of the war and was considered by many of those who operated it to be their best such weapon of the war. Its maximum range was 7680m (8400yds) with a standard 5.7kg (12.6lb) high explosive shell.

Three specialized tapered bore anti-tank guns were introduced during the war to fire tungsten

Now a museum piece, the German 75mm Pak 41 was the largest of three powerful anti-tank weapons that featured a tapered bore and fired a skirted projectile.

projectiles with a configuration known as the Gerlich system. The bores of the weapons narrowed from the breech to the muzzle, and a series of flanges folded inwards to allow the projectile to leave the weapon with high velocity. The 28mm sPzB 41, which fired a 0.125kg (4oz) projectile capable of penetrating 5.5cm (2.2in) of armour at 366m (400yds), was the first tapered bore gun produced. The second gun in the series was the 42mm lePak 41, which was mounted on the carriage of the old Pak 35/36, fired a 0.3kg (12oz) projectile that could penetrate 7.2cm (2.84in) of armour at a distance of 457m (500yds), and was issued to the Fallschïrmjager, or airborne troops. The shortage of tungsten ended the promising career of the heaviest tapered bore gun, the 75mm Pak 41. This weapon was capable of firing a 2.5kg (5.5lb) projectile through 17cm (6.73in) of armour at 457m (500yds). However, tungsten was a precious commodity in Nazi Germany, and the bulk of it was needed for machining tools. Thus, only 150 of the 75mm Pak 41 were produced and then silenced when their stocks of ammunition were depleted.

In 1936, the Skoda 47mm PUV vz 36, impressive with its ability to penetrate 5cm (2in) of armour at 640m (700yds) with a 1.5kg (3.3lb) high explosive shell, was authorized by the Czech Army. The weapon appeared to be obsolete, primarily due to its outdated spoked wheels, but by 1938 the Germans had pressed hundreds of the gun, both configured for mobility and for static use, into service as the 47mm Pak 36(t). Although its performance was soon eclipsed by heavier weapons, the gun remained in service with reserve units until the end of the war.

Early anti-tank guns

The earliest of the British anti-tank guns, the Ordnance, Q.F., 2-pounder, was developed primarily by Vickers-Armstrong in response to army specifications drafted in 1934. Five years later, the most widely used version of the gun, the Mark III, was delivered. For its firepower, the 2-pounder (47mm) was quite heavy and complicated to operate, although it must be said that the intent was for the weapon to be used in static defence rather than mobile warfare. During the campaign

in France in 1940 and the fighting in the desert of North Africa, it became painfully obvious that the weapon was inadequate to penetrate the armour of frontline German tanks. Its maximum range was only 549m (600yds), and its penetrating power was just over 5cm (2in) at 457m (500yds). While the weapon was withdrawn from the Royal Artillery by 1942 and reissued to infantry units, it remained in service in the China-Burma-India theatre, where it was still effective against lightly armoured Japanese tanks.

The successor to the 2-pounder was the Ordnance, Q.F., 6-pounder, which was placed in service in late 1941 following a series of production delays. The 6-pounder (57mm) improved the survivability of British gun crews with a penetrating power of 6.8cm (2.7in) at a distance of 914m (1000yds). Several variants of the weapon were

produced for anti-tank or tank mounted use. The Mark II and Mark IV were placed in the anti-tank role and distinguished by slightly different barrel lengths.

The US Army adopted the 6-pounder after realizing that its 37mm gun was inadequate, and redesignated it the 57mm Anti-tank Gun M1. The Soviet Union also received quantities of the 6-pounder. However, by 1943, the heavy German Tiger tank was being seen more often on the battlefield, and the 6-pounder was no match. It was withdrawn from the Royal Artillery that year but remained in service with the British Army.

The Ordnance, Q.F., 17-pounder was introduced as a prototype in August 1942, and rushed into service in North Africa on the carriage of the 25-pounder to address the threat of the Tiger tanks. Not until 1943, during the Italian

Q.F., 2PDR MK VII

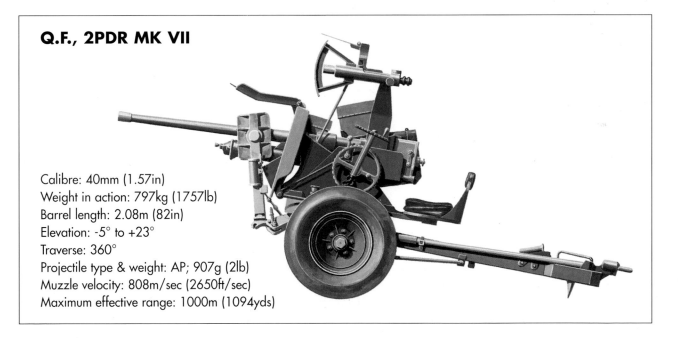

Calibre: 40mm (1.57in)
Weight in action: 797kg (1757lb)
Barrel length: 2.08m (82in)
Elevation: -5° to +23°
Traverse: 360°
Projectile type & weight: AP; 907g (2lb)
Muzzle velocity: 808m/sec (2650ft/sec)
Maximum effective range: 1000m (1094yds)

HEAD TO HEAD: *75mm Pak 40* VERSUS

Developed in late 1939 to counter a new generation of heavily armoured Soviet tanks, the 75mm Pak 40 was a heavier weapon than the previously deployed 50mm Pak 38. It was generally believed to be the finest German anti-tank weapon of World War II.

75mm Pak 40

Calibre: 75mm (2.95in)
Weight in action: 14425kg (3141lb)
Barrel length: 3.7m (12.1ft)
Elevation: -5° to +22°
Traverse: 65°
Projectile type & weight: APCR: 3.18kg (7lb); AP/HE: 6.8kg (15lb)
Muzzle velocity: APCR: 990m/sec (3248ft/sec); AP/HE: 792m/sec (2598ft/sec)
Maximum effective range: AP: 2000m (2190yds); HE shell: 7600m (8310yds)
Armour penetration: APCR: 154/500/90°; AP/HE: 132/500/90°

STRENGTHS

- High muzzle velocity
- Varied ammunition
- Excellent range

WEAKNESSES

- Heavy weight
- Limited crew protection
- Limited mobility

QF 6pdr 7cwt Gun Mk 2

The need for a heavier anti-tank weapon than the standard 2-pounder gun gave rise to the production of the 6-pounder in Great Britain during the autumn of 1941; however, delays resulted in more than a year passing before the weapon reached frontline troops.

QF 6pdr 7cwt Gun Mk 2

Calibre: 57mm (2.2in)
Weight in action: 1144kg (2521lb)
Barrel length: 2.56m (101in)
Elevation: -5° to +15°
Traverse: 90°
Projectile type & weight: AP: 2.72kg (6lb);
    APDS: 1.47kg (3.25lb)
Muzzle velocity: AP: 821m/sec (2695ft/sec);
    APDS: 1235m/sec (4050ft/sec)
Maximum effective range: 1500m (1650yds)
Armour penetration: AP: 74/1000/60°;
    APDS: 146/1000/60°

STRENGTHS

- Functional simplicity
- Good firepower
- Versatile configurations

WEAKNESSES

- Production delays
- Lack of heavy armour penetration
- Limited anti-tank role

Soviet infantrymen manning a 45mm anti-tank gun look on as a German shell explodes near their position.

was also used as the main armament for US light tanks and armoured cars. The 37mm cannon penetrated 2.5cm (1in) of armour at a distance of 914m (1000yds) and fired an armour-piercing projectile against hard targets.

The American 3in Model M5 was a combination of components of other ordnance, including the breech of the 105mm Howitzer M2A1 and the 76.2mm gun of the M3 anti-aircraft weapon. Introduced in December 1941, the gun was comparable in size and ease of deployment to others of the period. However, its weight of 2654kg (5850lb) was extreme, and large trucks were generally required for movement. Its 7kg (15.5lb) projectile was capable of

Armour and Survivability

The Achilles heel of the vaunted German Tiger tank was that its numbers were a relative few. More than 40,000 Sherman tanks were produced by the Allies, for example, while a mere 1,855 Tiger I and Tiger II tanks were produced in Germany. The high-velocity 88mm main gun of the Tiger was capable of destroying Allied Sherman or Churchill tanks from ranges greater than 1600m (1750 yds). In contrast, the earliest Shermans deployed to Europe, with their 75mm guns, could not penetrate the frontal armour of the Tiger at any range and needed to close within 500m (547yds) to penetrate the Tiger's side armour. Later, upgunned versions of the Sherman were more successful, but by then the Tiger had achieved almost mythical status among Allied tank and anti-tank artillery crews.

Campaign, did the 17-pounder (76.2mm) gun reach the battlefield in numbers. The weapon appeared to be unwieldy, particularly due to the length of its barrel, vertical breechblock and muzzle brake, but it was actually relatively simple to handle. It was serviced by a crew of seven, and its projectile could penetrate 13cm (5.12in) of armour plating from a distance of 914m (1000yds). The rate of fire for such a large gun was impressive, up to 10 rounds per minute. As the war neared its end, the gun had become standard among the Royal Artillery.

The US 37mm M3 earned the dubious distinction of being obsolete at the time it was issued to combat units headed to North Africa. It was withdrawn from the European theatre in favour of heavier types. However, the weapon was widely used in the Pacific, uniquely suited for use during amphibious operations and against light Japanese tanks. Firing a variety of shells, it was effective against enemy personnel and in the reduction of bunkers and pillboxes. Nearly 19,000 of the field variant were built during the war, and it

penetrating 8.4cm (3.31in) of armour from 1829m (2000yds). Due to the demands for the gun to arm the open turret M10 tank destroyer, the anti-tank weapon was slow to come off assembly lines. Only 2500 were completed during the war in the anti-tank configuration, while nearly three times that number were assembled for the M10.

The original Soviet 45mm anti-tank gun was identical to the Rheinmetall 37mm Pak 35/36 and was actually purchased from the German manufacturer in 1930. Two years later, the Soviets debuted their own version of the weapon upgunned to 45mm. Further revisions were completed in 1937, and a tank variety was mounted in 1938. While their performance was adequate during the Winter War with Finland in 1939–40, it was quickly realized that the gun could not stand up to heavy German armour. Although the need was recognized, Soviet industry was hard-pressed to come up with a new weapon before 1942. As it turned out, the M1942 was essentially the existing model with a longer barrel. A 57mm gun, the M1941, was in development, but the M1938 was switched from tank mounting to the rolling carriage as an emergency measure. The M1942, firing a 1.4kg (3.1lb) shell that could penetrate 9.5cm (3.74in) of armour at 302m (330yds), remained in service throughout the war.

The Soviet 76.2mm anti-tank weapon was originally introduced as a field gun and pressed into an anti-tank role. The Germans captured large numbers of the guns and adapted them for anti-tank use as the 76.2mm Pak 36(r). A later version, lighter than the prevalent M1936, was introduced as the M1939. However, the first 76.2mm gun intended for the dual field and anti-tank role was the M1942 with a muzzle brake and light carriage

Q.F., 17PDR
Calibre: 76.2mm (3in)
Weight in action: 2923kg (6444lb)
Gun Length: 3.562m (11ft 8in)
Elevation: -6° to +16.5°
Traverse: 60°

Muzzle velocity: 950m/sec (2900ft/sec)
Maximum effective range: 3000m (328yds)
Armour penetration: 130mm (5in) at 915m (1000 yards)

for ease of transport. In the final analysis, the M1942 is considered by many to be one of the most versatile weapons of the entire war. The gun was effective firing a projectile of up to 7.6kg (16.8lb) a distance of 13,337m (14,585yds). It was also capable of penetrating 9.8cm (3.86in) of armour at 498m (545yds).

During the 1930s, the Japanese produced the lightweight 37mm Type 94 and later the 37mm Type 97, a derivative of the German Pak 35/36. By 1941, however, it was realized that these weapons were inadequate against anything but the lightest of armoured vehicles. For this reason, the only Japanese-produced anti-tank gun of the war, the 47mm Type 1, was introduced. Light in comparison to anti-tank guns being developed in Europe, the Type 1 was nonetheless the heaviest such weapon available to the Japanese. In spite of the fact that it could never be produced in large

quantities, some were diverted to arm the Type 97 light tank.

Once the French government realized the quality of German armour that its armed forces were likely to face with a renewal of hostilities, the 47mm antichar [anti-tank] SA M1937 was rushed into production based on an Atelier de Puteaux design. Although it was never produced in adequate numbers, the majority of these were supplied to the French Army in 1939. Along with the mobile anti-tank model, a variant suitable for fixed fortifications was produced for the Maginot Line. The weapon was capable of penetrating 8cm (3.15in) of armour at 201m (220yds), more than a match for enemy tanks. A later version, the M1939, never saw combat due to the fall of France in the summer of 1940. The Germans appreciated the qualities of the M1937s that fell into their hands and put them into action as the 47mm Pak 141(f).

The French Canon de 47 antichar SA mle 1939 was an improvement over an earlier version of the anti-tank weapon. Its carriage provided 360-degree traverse.

Two light 25mm cannon, the antichar SA-L M1934 and the 25mm antichar SA-L M1937, were used by the French Army in World War II. The basic design of the M1934 had been conceived during World War I, but it did not reach the battlefield. In 1932, Hotchkiss designers responded to a request from the army by placing the weapon on a carriage, and the new configuration was adopted two years later. On the eve of World War II, more than 3000 had been issued. The M1937 was intended as an infantry support weapon rather than in an anti-tank role. With the outbreak of hostilities, the M1934 was briefly issued to soldiers of the British Expeditionary Force. Neither weapon performed well, capable of penetrating only 4cm (1.57in) of armour at 440 yards with a maximum range of just under 1828m (2000yds). Those which fell into German hands were apparently retired by 1942.

Keeping up with the armour

The often fluid conditions of the modern battlefield necessitated the development of artillery that could not only be deployed easily but was also capable of moving under its own power. Tanks had revolutionized warfare, and it became incumbent upon infantry and artillery to keep pace with armoured spearheads.

Author Chris Bishop explains:

'Self-propelled artillery was very much a product of the type of warfare that evolved during World War II: before 1939 self-propelled artillery scarcely existed (apart from a few trial weapons), but by 1943 it was used by all the combatant nations. The sudden rise of this new form of weapon can be attributed almost entirely to the impact of the battle tank on tactics, for warfare no longer took place at the speed of the marching soldier and the scouting horse, but at the speed of the tank. These swarmed all over Poland, France and eventually the Soviet Union, and the only way that the supporting arms, including the artillery, could keep up with them was to become equally mobile.'

Early self-propelled artillery examples consisted largely of the combination of existing tank chassis and proven guns of the artillery. So it was with the leFH 18/2 auf Fgst Kpfw II(Sf) SdKfz 124 Wespe, or simply the Wespe. A gem of a self-propelled artillery piece, the Wespe (Wasp) was proposed in 1939 to provide mobile artillery to frontline units, by utilizing the dependable but outdated hull of the Panzerkampfwagen II tank. In converting the tank to a self-propelled gun, the turret was redesigned to be open for the crew of five, and a raised armour shield offered a degree of protection. The gun was initially equipped with a 105mm leFH 18 howitzer and a 7.92mm machine gun. From 1940 to 1944, nearly 700 of the weapons were produced, primarily at factories in Poland.

The Wespe was first committed to action on the Eastern Front in 1943 and deployed in batteries of six guns each. Reportedly, the Wespe performed so well that other systems utilizing the PzKpfw II chassis were discontinued.

The predecessor of the Wespe was the sIG 33 auf Geschützwagen, armed with a 150mm howitzer mounted on the chassis of the PzKpfw I tank. First appearing with field units during the 1940 conquest of France, the sIG 33 was conspicuous with its high profile and open turret with room for a crew of four. In 1942, versions utilizing the PzKpfw II and PzKpfw 38(t) chassis were introduced with the howitzer mounted lower in the turret. The 38(t) version, known as the Bison, proved quite satisfactory and resulted in a later series, the SdKfz 138/1, which was produced in total rather than from converted tank hulls and remained in production until the end of the war. A 1941 attempt to utilize the PzKpfw III chassis was short-lived. Nearly 400 of the Bison and SdKfz 138/1 were built during the war.

Perhaps the most recognizable of the German self-propelled artillery was the Sturmgeschütz III, originally developed in the late 1930s with a short 75mm gun mounted on the PzKpfw III chassis. A series of revisions was concluded with the G variant, and the short 75 was replaced with a long-barrelled weapon in later models. The Sturmgeschütz III performed better in an infantry support role than as an anti-tank weapon, although the long barrel 75mm gun did provide such capability. Secondary armament included two 7.92mm machine guns, and the vehicle was operated by a crew of four. The gun was distinctive for its low profile and for armoured plating that was fastened a slight distance from the hull to detonate shells before they came in contact with the crew compartment. Near the end of the war, some Sturmgeschütz IIIs were upgunned to a 105mm howitzer. A somewhat redesigned version, the Sturmgeschütz IV, was also produced.

Concerns over the armoured protection of the Sturmgeschütz III arose in 1942, and it was further believed that the armour protection afforded by the sIG 33 was inferior to many of the new anti-tank weapons fielded by Allied troops. Therefore, in the summer of 1943, a more heavily armoured self-propelled artillery piece, carrying a 150mm howitzer and two 7.92mm machine guns, entered service. Designated the Sturmpanzer IV Brummbär (Grizzly Bear), the vehicle was also routinely coated

Left: The largest German self-propelled weapon intended for the close support of ground troops was the Sturmtiger, which fired a modified naval depth charge.

with a paste called zimmerit, which prevented magnetic mines from being attached to the hull. More than 300 of the weapon were produced and provided direct support, particularly during close combat, to panzergrenadier units. The husky Brummbär weighed more than 28,123kg (62,000lb), and its frontal armour was more than 6.3cm (2.5in) thick.

Mounting a 150mm howitzer and a single 7.92mm machine gun, the self-propelled artillery version of the 24,000kg (52,911lb) Hummel (Hummingbird) was an imposing sight on the battlefield, as was the 88mm-gunned anti-tank version of the weapon. Both were constructed atop a hybrid of the PzKpfw III and IV chassis, and some were built without the howitzer as ammunition carriers. The 150mm version of the weapon served as the mobile artillery support for panzer and panzergrenadier divisions of the German Army after 1942. It was crewed by five men, and nearly 700 were produced by the end of the war.

Harkening back to the days of gigantic siege mortars a generation earlier, the Karl series was produced in the late 1930s. According to some scholars, the Karl series constitutes the largest self-propelled artillery ever constructed. Weighing up to 124,000kg (273,373lb), the mortar was produced in 540mm or 600mm variants. The range of the former was 6240m (6824yds), while the latter was 4500m (4921yds). They hurled shells of up to 1250kg (2756lb) and 2170kg (4784lb) respectively and were capable of penetrating up to 3.5m

(11.5ft) of concrete prior to detonation. While the Karl mortars were technically self-propelled, manoeuvrability was understandably limited. Travel of any distance required transport by train or disassembly of the barrel and movement aboard trailers towed by tractors.

Designed with the reduction of large fortifications in mind, the Karl series was intended first for use against the Maginot Line. However, France fell before they were deployed, and their first shots were fired in anger on the Eastern Front against the Brest fortress. Later, the Karl was used during the siege of Sevastopol in the Crimea. Six

Karl series guns were ultimately produced, and they were named Adam, Eva, Thor, Loki, Ziu and Odin. Ziu was fired during the Warsaw Uprising of 1944.

Two additional German self-propelled artillery pieces are worthy of mention. Following the experience of the German Army at Stalingrad, the decision was made to rearm the chassis of the Tiger I heavy tank with a 210mm (8.26in) mortar for fighting in crowded urban areas. Difficulty obtaining this gun prompted the installation of a 380mm (14.9in) mortar rocket launcher, and the new weapon was named the Sturmtiger or Sturmpanzer VI. The modified vehicle weighed

BISHOP

Crew: 4
Weight: 7879kg (17,333lb)
Dimensions: length: 5.64m (18.5ft); width: 2.77m
  (9.09ft); height: 3.05m (10ft)
Range: 177km (110 miles)
Armour: 8–60mm (0.315–2.36in)

Armament: one 25-pounder (87.6mm/3.45in) gun
Powerplant: one AEC six-cylinder diesel engine
  developing 98kW (131hp)
Performance: maximum road speed: 24km/h
  (15mph); fording: 0.91m (3ft); vertical obstacle:
  0.83m (2.7ft); trench: 2.28m (7.5ft)

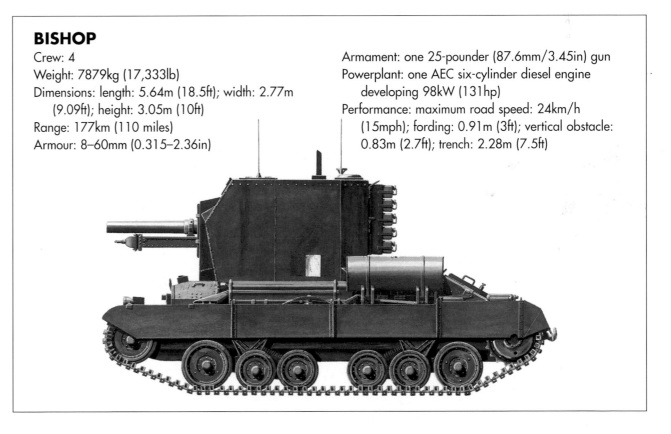

61.6 tonnes (68 tons) and the projectile weighed 345kg (761lb). Three panzer companies were recruited to operate the Sturmtiger, which was used during the Warsaw Uprising, the Battle of the Bulge and the fighting in northern Italy. The second system, the Waffentrager, amounted to a self-propelled transporter that carried an artillery piece into action but did not always serve as a firing platform. The system was reported to normally carry a 105mm howitzer, and eight were known to have been built.

The notable self-propelled artillery of the British Commonwealth included the Sexton and the Bishop. The Sexton amounted to a 25-pounder cannon fitted into an open turret atop the Ram tank chassis. The Sexton was similar in design to the American M7 Priest but did not have such a prominent pulpit configuration. Employed as a field artillery weapon almost exclusively, it was served by a crew of six. Secondary armament included one Browning .50-calibre machine gun and two unmounted 7.7mm Bren guns. The Bishop, the British Army's first self-propelled artillery, was the combination of the 25-pounder gun and the Valentine tank chassis, conceived in order to provide more protection for the gun crew. First appearing in service in North Africa, the Bishop was also used during the Italian Campaign. Its deficiencies included cramped crew space and limited elevation capability. Late in the war, most of the Bishops were abandoned for the American M7 Priest.

Mounted on the chassis of the M3 medium tank, which was fast becoming obsolescent, the M7 Priest self-propelled gun owed much of its success to the experience of the Americans with 75mm and 105mm howitzers mounted on halftracks. The Priest received its nickname reportedly from British

Red Army Juggernaut

As the Soviets advanced towards Berlin, they relied on the superb performance of armoured vehicles, including the T-34 medium tank. However, self-propelled assault guns played a key role in the great victories achieved on the Eastern Front in 1944 and 1945. The pair of ISU-152 assault vehicles pictured carried the formidable 152mm howitzer.

soldiers who noted the height of its .50-calibre machine-gun position. The Priest was used in support of US troops in Europe and the Pacific, and went to war with the British at El Alamein in North Africa during the autumn of 1942. As the war progressed, production of the M7 was sporadic, and a later variant was configured with the chassis of the M4 Sherman tank. Unarmed

personnel carriers using the M7 configuration were nicknamed Kangaroos.

In late 1943, the 155mm M40 went into service with the US Army as a replacement for the M12, the country's first 155mm self-propelled gun. The M12 had performed well but was in need of updating, and its replacement was the M40, which combined the 155mm M1A1 Long Tom cannon with the chassis of the M4A3 Sherman tank. The first M40s did not appear in combat until 1945 and participated in the bombardment of German cities such as Cologne. More than 300 M40s were built during the early months of 1945, and production continued after the end of the war. The M43, mounting a 203mm howitzer, was also placed in production, but less than 50 were completed.

The Soviet Union deployed the ISU-152 and ISU-122 fully enclosed assault guns in the spring of 1943, and the vehicles were virtually identical with the exception of their 152mm and 122mm main armaments. The two guns were housed in mantlets atop the chassis of the KV II heavy tank and later that of the IS tank. As was typical of Soviet weaponry, the dual-purpose guns were equally adept in the assault or anti-tank role. Each carried secondary armament of one .50-calibre machine gun, and its silhouette was characteristic of Soviet armour, shunted forwards. The SU-76, which employed the ubiquitous 76.2mm gun and the chassis of the discontinued T-70 light tank, arrived on the Eastern Front in mid-1943, just as a new generation of German tanks was arriving as well. When it became apparent that the 76.2mm gun was not faring as well against enemy armour as it had in the past, the weapon was refocused to provide fire support for Red Army infantry. Secondary armament consisted of one 7.62mm machine gun.

Just as they lagged in development of artillery and tanks, the Japanese produced very few self-propelled guns during World War II. The most notable of these was the Type 4 HO-RO, which combined the Type 38 150mm howitzer, dating back to a 1905 Krupp design, and the chassis of the Type 97 tank. Lightly armoured, the Type 4 was still constructed by riveting, which had long been replaced by welding in other countries, and was manned by a crew of either four or five. Another self-propelled weapon, the Type 2, mounted a 75mm gun and was also below average in performance.

The wail of Stalin's organ

To the veterans, the mournful sound was unmistakable. At Stalingrad, a salvo of Soviet 132mm rockets killed an entire battalion of German soldiers in a blinding flash. Now, as the Red Army closed in on Berlin, the German capital, the Katyusha was a messenger of destruction and a harbinger of death.

In *Battlefield Berlin*, Peter Slowe and Richard Woods write:

'Troops started pouring into Beeskow, as Zhukov's and Koniev's armies advanced and the local people came out into the streets to watch as half the Ninth Army wound its way through, nearly 100,000 men, some apparently in good order and

The ISU-122 self-propelled assault weapon mounted the versatile 122mm cannon and complemented the somewhat heavier ISU-152, which mounted a 152mm gun.

HEAD TO HEAD: *Wespe* VERSUS

The leFH 18/2 auf Fgst Kpfw II (Sf) SdKfz 124 Wespe, or Wasp, was a highly successful artillery support vehicle, mounting a 105mm howitzer and an MG34 machine gun atop a PzKpfw chassis. The weapon first appeared in service in 1942.

Wespe

Crew: 5

Weight: 11,000kg (24,251lb)

Dimensions: length: 4.81m (15.8ft); width: 2.28m (7.5ft); height 2.3m (7.5ft)

Range: 220km (137miles)

Armour: (Ausf version): 20–35mm (0.8–1.38in)

Armament: one 105mm (4.13in) howitzer with 32 ready-use rounds and one 7.92mm (0.31in) machine gun

Powerplant: one Maybach six-cylinder petrol engine devleoping 104kW (140hp)

Performance: maximum road speed: 40km/h (25mph); fording: 0.85m (2.8ft); vertical obstacle: 0.42m (1.4ft); trench 1.75m (5.7ft)

STRENGTHS

• Compact configuration
• Reliability
• Excellent mobility

WEAKNESSES

• High profile
• Open fighting compartment
• Interrupted production

M7 Priest

The M7 Priest presented one of the most recognizable silhouettes of any armoured vehicle produced during World War II. Its 105mm howitzer was mounted atop the chassis of an M3 medium tank along with a single .50-calibre machine gun. The vehicle went into action in 1942 with British forces at El Alamein.

M7 Priest

Crew: 5
Weight: 22,500kg (49,500lb)
Dimensions: length: 6.02m (19.8ft); width: 2.88m
    (9.4ft); height: 2.5m (8.2ft)
Range: 201km (125 miles)
Armour: UP TP 25.4mm (1in)
Armament: one 105mm (4.1in) Howitzer
Powerplant: one Continental 280kW (375hp)
    nine-cylinder radial piston engine
Performance: maximum speed: 42km/h (26mph);
    fording: 1.22m (4ft); vertical obstacle: 0.61m (2ft);
    trench: 1.91m (6.3ft)

STRENGTHS

• Availability
• Adequate crew space
• Reliability

WEAKNESSES

• Prominent silhouette
• Open fighting compartment
• Thin armour protection

The M-13 132mm rocket, known far and wide as the Katyusha, was one of several rocket systems employed by the Red Army during World War II, but it was the most widely used. Only a few of the weapons were available when the Nazis launched Operation Barbarossa on 22 June 1941, and invaded the Soviet Union. A month after the invasion, when just about everything else was going wrong for the Soviets, the rockets sruck panic into German soldiers at Smolensk.

With a range of 8500m (9295yds), the 132mm fin-stabilized rocket weighed 42.5kg (93.7lb) and delivered 4.9kg (10.8lb) of explosive. The launchers, designated BM-13-16, were carried on ZiS-6 6X6 trucks. Although it was not aimed, the Katyusha was fired in salvoes of as many as 16 per vehicle. Soviet authorities attempted to maintain secrecy as to the components of the rocket launch system, but to no avail. Lend Lease trucks from the United States were also modified to carry the launchers, and several types of rockets were developed during the course of the war. One system, the M-13-DD, was powered by two rockets and had the greatest range of any solid-fuelled rocket weapon of the war at 11,800m (12,905yds). The M-13 remains in use around the world today, although it was replaced in Soviet arsenals during the 1980s.

Other Soviet rocket systems of World War II were also generally known by the Katyusha moniker, including the M-8 82mm rocket, which was launched in salvoes of up to 24 from both trucks and the modified hulls of T-60 and T-70 light tanks. With a range of 5898m (6450yds), the rocket delivered a 0.5kg (1.1lb) fragmentation warhead, which was capable of inflicting heavy casualties on massed infantry.

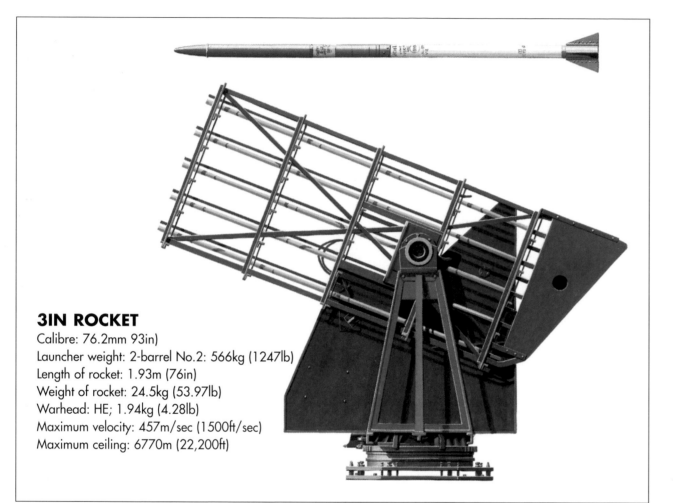

3IN ROCKET

Calibre: 76.2mm 93in)
Launcher weight: 2-barrel No.2: 566kg (1247lb)
Length of rocket: 1.93m (76in)
Weight of rocket: 24.5kg (53.97lb)
Warhead: HE; 1.94kg (4.28lb)
Maximum velocity: 457m/sec (1500ft/sec)
Maximum ceiling: 6770m (22,200ft)

good heart, most in rapid retreat mixed up with civilian refugees. From the east of the town, they suddenly heard a loud whining noise. There were no planes, and it was not a bombing raid. Some soldiers who were in the know dived for cover. For a moment the locals seemed mesmerized, staring in the direction of the noise. All at once it was too late. Thousands upon thousands of shells exploded in Beeskow. Crowds of townspeople became piles of torn flesh and flowing blood as they were blown up. A cattle shed was hit, and the scene was made more hideous as the maddened beasts, some of them on fire, charged through the exploding streets. The whining noise had been the music of Stalin's organs, Katyusha rockets playing on the retreating Ninth Army. Beeskow had become a front-line town.'

A British soldier kneels to fire the LILO, which was a short-range rocket introduced late in World War II for use against hardened strongpoints, such as Japanese bunkers.

30 prevented mobile launching. After the war, both models were discontinued in favour of weapons with longer range.

Early British endeavours in rocket design were concentrated primarily on anti-aircraft applications, and a mildly successful 2in rocket was produced in the late 1930s. Often mounted on naval vessels and merchantmen, the 2in rocket was armed in flight by a wind vane and set to detonate 4.5 seconds after launch. It was also designed to carry a light wire aloft, theoretically tangling in the propellers of low-flying enemy planes and causing them to crash. Other naval and land versions of the rocket were the subject of experimentation, but these were abandoned for larger weapons. At just over 0.225kg (8oz), the warhead of the rocket was not substantial.

In 1939, test firings of a 3in rocket were conducted in Jamaica. Still intended as an anti-aircraft weapon, the 3in warhead weighed 1.9kg (4.2lb). Several revisions were conducted over the next three years, and a few mobile multiple launchers, capable of firing up to 36 rockets in four rippled salvoes, were deployed in North Africa. In 1944, a system capable of firing 20 rockets in rippled salvoes was placed in home defence service. Unexpectedly, the 3in rocket proved to be an outstanding aerial weapon, fired at German armour from short rails under the wings of Hawker Typhoon and Tempest fighter-bombers.

In 1944, the LILO rocket system was introduced with the intention of coping with Japanese bunkers and gun emplacements. The LILO was notoriously

The heavy M-30 300mm rocket system was deployed by the Soviets in 1942, delivering 28.9kg (63.7lb) of explosives more than 2743m (3000yds). These were fired from carrying crates called Rama, which were heavy and difficult to set into position.

Later in the same year, the M-31 debuted, increasing the range to 4302m (4705yds) with an improved rocket motor. The M-31 was made mobile with the introduction of truck-mounted launchers in 1944, but the limited range of the M-

inaccurate but fired a warhead of up to 27.2kg (60lb), which was capable of penetrating 3m (10ft) of ground before detonating. The rocket was fired electrically and normally serviced by two men, one who carried the launcher and another with the rocket itself.

A hybrid rocket system known as the Land Mattress was composed of a 12.7cm (5in) naval warhead and a rocket motor originally used with aircraft. It was placed in service with the army in 1944 and performed well during the fighting around the Scheldt Estuary near the port city of Antwerp, Belgium. Towed launchers for the Land Mattress accommodated up to 32 of the 7.6cm (3in) rocket. The weapon's main shortcomings were limited range due to minimum and maximum elevations of only 23° and 45°. Capable of firing up to 7224m (7900yds), the Land Mattress would overshoot targets less than 6127m (6700yds) distant. Despite its success, the Land Mattress was produced only in limited numbers before the war ended. However, it was reported that the system could place half the rockets fired in a single salvo within a relatively small area of 210m (230yds) square.

German rocket systems were originally intended to generate smoke, obscuring troop movements. However, they quickly found purpose in delivering high explosives. To the Allied infantryman, the word nebelwerfer (smoke thrower) came to describe all German rockets. The dull drone made by in flight also resulted in the nickname 'Moaning Minnies'.

A veteran of the New Zealand 5th Field Regiment, Owen Raskin remembers close calls with German rockets:

'We also became acquainted with nebelwerfers. They were really scary at all times, but especially so

A Soviet M13 truck-mounted rocket launcher fires its complement of rockets amid the rubble of a war-ravaged street.

at night. Familiarity, however, reduced their effect to some degree. When they were fired, the sound was like a chorus of sighing groans and screeches, followed by silence while the bombs were in flight and everyone speculated as to where they would land. Les Morrison says that he and two or three others were … chatting and waiting for nebelwerfers to land. The silence ended with a bomb landing two or three feet in front of the gun. No one was hurt but they all went to ground after the explosion. They had not heard it coming. On another occasion a projectile landed just outside B Troop command post. One landed on a bivvy tent

but fortunately no one was in it. Les and others described the missile casing as being of light metal that curled back like a half-peeled banana on exploding. It was suggested that apart from putting the wind up all who felt they might be the target, the nebelwerfer was intended to kill or injure by blast rather than fragmentation. Nevertheless a small piece of shrap from an explosion in the E Troop area about 150 yards away knocked my writing material from the ground of my sleeping bivvy onto my head. Next morning its track through a paddock of young oats could be seen in line with a hole in my tent.'

The Germans deployed the 150mm Wurfgranate 41 and the 210mm Wurfgranate 42, launched from the multi-barrelled nebelwerfers mounted on the carriage of the Pak 35/36 gun or fixed atop the SdKfz 4/1 halftrack. The 150mm is probably the rightful owner of the Moaning Minnie nickname, and its 2.5kg (5.5lb) warhead could reach targets more than 6858m (7500yds) distant. The 210mm rocket contained a 10kg (22.4lb) high explosive warhead and ranged to nearly 7864m (8600yds).

Preceding the 150mm and the 210mm rockets were the 280mm and 320mm Wurfkörper, which were fired from wooden or steel launching frames that doubled as carrying cases. To add mobility to these early rockets, a towed launcher with six frames was constructed. Later, mobile launchers were mounted on halftracks and other vehicles. These rockets were devastating when they struck a target and were used to good advantage in urban combat. They continued in service throughout the war.

In 1942, the 300mm Wurfkörper 42 went into service and was initially fired from the 300mm Nebelwerfer 42 launcher. Among the improvements of the Wurfkörper 42 were a propellant that left a much diminished smoke trail, minimizing the risk of return fire from enemy artillery batteries. Later, a trailer based on the Pak 38 anti-tank gun was introduced, and the weapon could be launched from a halftrack mount as well. The 44.6kg (98.4lb) explosive charge carried by the rocket was large compared to the overall projectile weight of 125.6kg (277lb).

Japanese rocket development lagged behind that of the other warring nations, and different research teams often worked in opposition to one another. The army and navy each developed 200mm rockets, the army version being designed for use with the Type 4 launcher, which was essentially a large mortar. The navy 200mm rocket was launched from wooden or metal rails or sometimes even planted in the ground. Other Japanese rockets included the large 447mm weapon used in the Philippines and on Iwo Jima along with the Type 10 rocket motor, which was attached to a modified aerial bomb, propelling it down a launch ramp.

WURFGRANATE 41

Crew: 3
Weight: 7100kg (15,653lb)
Dimensions: length: 6m (19.7ft); width: 2.2m (7ft);
    height: 3.05m (10ft)
Range: 130km (81 miles)
Armour: 8–10mm (0.31–0.39in)

Armament: 1 x 150mm (5.9in) Nebelwerfer 41
    10-barrelled rocket launcher, 1 x 7.92mm
    (0.31in) MG
Powerplant: 1 x Opel 6-cylinder petrol
Performance: maximum road speed: 40km/h
    (25mph)

More than 2.5 million of the American 4.5in M8 rocket were produced during the war, even though the United States had virtually no rockets in service and had conducted little research in 1941. A variety of truck mounted and towed launching systems fired up to 60 M8 rockets in a single barrage. The most recognized of these was the calliope, which was mounted above the turret of the M4 Sherman tank. The T44 launcher and the larger Scorpion were placed on amphibious DUKW craft and capable of launching 120 and 144 M8 rockets respectively. These were used frequently during the latter stages of the Pacific War. The M12 launcher fired a single M8 rocket against enemy bunkers and pillboxes. As research progressed, the spin stabilized M16 rocket was developed along with the T66 launcher, which could fire 24 rockets in two seconds. The maximum range of the M8 rocket was 4206m (4600yds), and its explosive warhead weighed 1.9kg (4.2lb).

Giants in the land

'The nearness of the previous night's shelling gave me much food for thought the next morning,' remembered V.C. Fairfield of the 64th Field Regiment, Royal Artillery.

'Later in the day I set about improving the safety of our command post which felt more vulnerable each time the enemy had a go at us… That night I was off duty for a few hours and slept in my personal slit trench which was as narrow as I could bear, about two feet deep but warm enough… However, the trouble with sleeping alone was that during the shelling I tended to develop an uncontrollable tremble… One member of the battery had had a "feeling" and had spent the night with the light anti-aircraft guns. On returning to his bivouac he found it wasn't there! Instead there was a large crater made by a 210mm shell!'

The experiences of Fairfield and his fellow artilleryman were not unique, particularly at the Anzio beachhead where the incident occurred. However, the 210mm shell was not the largest hurled at the beachhead. During the protracted

Left: A German soldier places camouflage over the launcher of a 300mm Wurfkörper 42 rocket system. The carriage was modified from that of the 50mm Pak 38 gun.

Right: The American T-34 60-tube rocket launcher, known as the calliope, was often mounted atop the chassis of the M4 Sherman tank. The tubes were made of plywood.

HEAD TO HEAD: *Katyusha* VERSUS

The truck mounted Soviet M-13 Katyusha rocket system was the most famous of World War II. Deployed in large numbers across the Eastern Front, the Katyusha gained the grudging respect of its Axis adversaries. The 132mm rockets were usually fired in mass barrages.

Katyusha

Crew: 2
Weight: 8900kg (19,600lb)
Dimensions: length: 6.55m (21.5ft); width: 2.24m
   (7.3ft); height: 2.76m (9ft)
Range: 370km (230 miles)
Armour: Not applicable
Armament: 16 x 132mm (5.2in) M-13 rockets
Powerplant: 1 x Hercules JXD 6-cylinder petrol,
   developing 65kW (87hp)
Performance: maximum road speed: 72km/h (45mph)

STRENGTHS

• Heavy firepower
• Excellent mobility
• Ease of deployment

WEAKNESSES

• Noticeable firing signature
• Lack of accuracy
• Limited elevation

28/32cm Wurfkörper sdKfz 251

Heavy but short-range 280mm or 320mm rockets collectively known as Wurfkörper were fitted to tracked vehicles by the Germans. Often, the mount was the SdKfz 251, a workhorse of the German armed forces. Such vehicles were nicknamed the Howling Cow by German soldiers.

28/32cm Wurfkörper SdKfz 251

Crew: up to 12
Weight: 8000kg (17,600lb)
Dimensions: length: 5.8m (19ft); width: 2.1m (6.9ft); height: 1.75m (5.7ft)
Range: 300km (190 miles)
Armour: 15mm (0.59in)
Armament: 6 x 28cm (11in) or 32cm (12.6in) Wurfkörper rockets; 1 x MG
Powerplant: 1 x Maybach HL 42 6-cylinder petrol, developing 75kW (101hp) engine
Performance: maximum speed: 53km/h (33mph); fording: 0.6m (2ft); gradient 24 per cent

STRENGTHS

- Heavy firepower
- Mobility
- Effective against structures

WEAKNESSES

- Limited range
- Highly inaccurate
- Bulky design

struggle for the once-beautiful resort town on the Tyrrhenian coast of Italy, the Germans brought forward a pair of heavy railway guns, which they had named Robert and Leopold. These guns were designated the K5 series and were the result of a research programme initiated by Krupp in the 1930s to produce heavy guns for support of infantry operations.

The K5s, which Allied troops collectively nicknamed 'Anzio Annie', were 283mm weapons with a range of 61m (38 miles) and a rate of fire of one round every three to five minutes. Test-firing occurred in 1936, and as many as eight of them entered service during the campaign in France in

A German 210mm K12 (E) railway gun fires its shell towards Allied positions on the Anzio beachhead. Also in use were two 283mm K5 guns, nicknamed 'Anzio Annie'.

1940. Later in the war, experiments were conducted with the K5 firing a rocket-propelled projectile for increased range. The two guns in action at Anzio were captured by Allied troops in June 1944. Both had been damaged; however, they were sent to the United States for testing. One complete gun, assembled from the parts of the two damaged weapons, remains on display at the US Army Ordnance Museum. The Soviets also deployed numerous large-calibre railroad guns produced during the 1920s and 30s.

The K5 guns were by no means the largest in the Nazi arsenal. Krupp also designed the mammoth 800mm Schwerer Gustav and Dora guns, named for the head of the armaments firm and the wife of the project's senior engineer, as weapons capable of destroying the French Maginot Line fortifications. Weighing nearly 1219 tonnes (1344 tons), the two guns were capable of firing a shell weighing 6.3 tonnes (7 tons) a distance of 59.5km (37 miles). The largest calibre rifled weapon in history, the Schwerer Gustav was exceeded in calibre by only a few large mortars. One of these was Little David, an American 914mm mortar originally intended for test-firing bombs. World War II ended before Little David was deployed to the Pacific.

The Schwerer Gustav was test-fired in 1939 and placed in service in late 1941. In June 1942, the weapon fired 48 rounds during the siege of Sevastopol. It was then transported to Leningrad to participate in an attack, which was subsequently cancelled, and returned to Germany. It was destroyed there in the spring of 1945 to prevent its capture. Dora participated in the Battle of Stalingrad but was also destroyed near the end of the war.

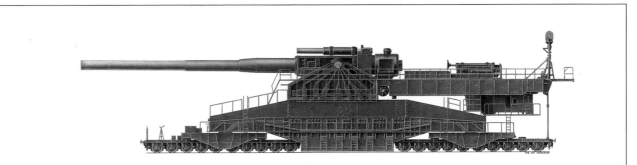

80CM SCHWERER GUSTAV

Calibre: 80cm (31.4in)
Barrel length: 32.48m (106.56ft)
Length over buffers: 42.98m (141ft)
Weight in action: 1,350,000kg (1329 tons)
Elevation: +10° to +65°
Traverse: nil

Shell type & weight: HE: 4800kg (4.73 tons); CP 7100kg (7 tons)
Muzzle velocity: HE: 820m/sec (2690ft/sec); CP: 710m/sec (2330ft/sec)
Maximum range: HE: 47km (29.2 miles); CP: 38km (23.6 miles)

Other giant guns constructed by the Germans included the 520mm Langer Gustav and the mysterious V-3 cannon. The Langer Gustav was intended to fire rocket-propelled shells of 680kg (1499lb) up to 190km (118 miles). It was never completed due to damage to its production facility from Royal Air Force bombing. The designation V-3 reportedly referred to a complex of five underground shafts located in France and housing gargantuan guns, which were potentially capable of hitting London. The V-3 was also supposedly one of Hitler's 'vengeance weapons', intended to inflict casualties on civilian targets in retribution for reverses suffered in battle. Reportedly, two of the guns fired more than 180 shells at targets in Luxembourg during a six-week period in early 1945. The

V-3 site and several disassembled guns are said to have been captured by American troops, shipped to the United States for testing, and destroyed in 1948.

Brave new world

With the defeat of Germany and Japan in 1945, the relationship between the United States and Great Britain in the West and the Soviet Union in the East, began to deteriorate. As political tensions and distrust brought about the half-century of discord that came to known as the Cold War, leaders and nations grappled with new military realities. The destruction of Hiroshima and Nagasaki had ushered in the atomic age. War was now a means by which mankind could annihilate itself.

THE COLD WAR

'The enemy of my enemy is my friend'. So says a centuries-old proverb of Eastern Europe. From time to time, political ideology has given way to military necessity – this was seen during World War II. The United States, Great Britain and the Soviet Union – a democracy, a constitutional monarchy and a totalitarian communist state – forged what could be described as a tenuous alliance.

With the defeat of Hitler and Nazi Germany, the frayed marriage of convenience took a decided chill, East and West growing more wary about the intentions of the other and the prospects for peaceful coexistence in the post-war world. Even before the end of hostilities in May 1945, the Big Three were polarizing, redrawing the map of Europe and asserting authority over spheres of influence. The vestiges of goodwill fell systematically away.

On 5 March 1946, Winston Churchill spoke at Westminster College in Fulton, Missouri. He warned:

'From Stettin in the Baltic to Trieste in the Adriatic an iron curtain has descended across the Continent. Behind that line lie all the capitals of the ancient states of Central and Eastern Europe. Warsaw, Berlin, Prague, Vienna, Budapest,

Left: US soldiers fire a recoilless rifle in Korea. Capable of firing heavier projectiles than recoiling weapons, these were popular until supplanted by anti-tank missiles.

Belgrade, Bucharest and Sofia; all these famous cities and the populations around them lie in what I must call the Soviet sphere, and all are subject, in one form or another, not only to Soviet influence but to a very high and in some cases increasing measure of control from Moscow.'

Scarcely 10 months had passed since the surrender of Nazi Germany. Churchill's speech is generally considered to mark the onset of the Cold War, half a century of intrigue, sabre-rattling, political brinkmanship and proxy wars. In the wake of the costliest and most dreadful armed conflict in human history, the world entered the atomic age, and the threat of absolute annihilation became real – and, at the same time, surreal. The nature of both military strategy and tactics had been altered. With the advent of the North Atlantic Treaty Organization (NATO) and the Warsaw Pact, Europe and, indeed, the world were once again divided into something resembling two armed camps. The doctrine of Mutually Assured Destruction evolved, and an arms race of

unprecedented proportion ensued. The term 'superpower' became a part of everyday lexicon.

During the Cold War, the technology of artillery, from delivery systems to destructive potential, increased tremendously. A new generation of systems capable of delivering tactical nuclear weapons broadened the battlefield and the potential price to be extracted in the event of open hostilities between the Western nations and the Soviet Bloc. As authors Richard A. Gabriel and Karen S. Metz write:

'Only eight years after Hiroshima, nuclear artillery shells were invented, and three years later these shells were small enough to be fired from a 155mm howitzer.., In 1980, the US Army estimated that modern non-nuclear conventional war had become 400 to 700 per cent more lethal and intense as it had been in World War II depending, of course, on the battle scenario. The increases in conventional killing power have been enormous, and far greater and more rapid than in any other period in man's history. The artillery

HEAD TO HEAD: *British L118 Light Gun* VERSUS

Designed during the late 1960s for the British Army and deployed in the 1970s, The L118 Light Gun was subsequently exported to a number of countries, including the United States. The L118 replaced the British version of the Italian-made OTO Melara Mod 56 as their standard light infantry support weapon.

British L118 Light Gun

Calibre: 105mm (4.13in)
Weight: (in travelling and firing orders) 1860kg
   (4100lb); (elevating mass) 1066kg (2350lb)
Barrel length: 3.175m (10.4ft)
Elevation: -5.5° to +70°
Traverse: (total on carriage) 11°; (total on platform)
   360°
Shell type & weight: HE L31: 16.1kg (35.5lb);
   HESH L42: 10.49kg (23lb); L37L, L38 and L45
   Smoke: 15.88kg (35lb); L43 Illuminating

STRENGTHS

• Excellent mobility
• Good range
• Combat tested

WEAKNESSES

• Needed ammunition upgrade
• Durability
• Complex technology

Mod 56

Developed in Italy during the 1950s, the OTO Melara Mod 56 105mm gun was designed as a light artillery piece that could be used by Alpini troops in rugged, mountainous terrain. The weapon remained in service for nearly 50 years after it was introduced.

Mod 56

Calibre: 105mm (4.13in)
Weight: 1273kg (2806lb)
Length: 14 calibre: 1.47m (57.9in)
Elevation: -7° to +65°
Traverse: 56°
Muzzle velocity: 416m/s (1345ft/s)
Maximum range: 11,100m (12,140yds)

STRENGTHS

- Light weight
- Easy assembly and disassembly
- Good firepower

WEAKNESSES

- Comparatively limited range
- Primarily alpine or airborne use
- Limited crew protection

firepower of a manoeuvre battalion, for example, has doubled since World War II while the "casualty effect" of modern artillery guns has increased 400 per cent. Range has increased, on average by 60 per cent, and the "zone of destruction" of battalion artillery by 350 per cent. A single round from an 8-inch gun has the same explosive power as a World War II 250 pound bomb.'

Frontline in Korea

One of the earliest flashpoints of the Cold War was the Korean peninsula. Although research into nuclear artillery and updated conventional weaponry was well underway, virtually all of the artillery engaged by the United Nations forces, as well as those of the communist North Koreans and later the People's Republic of China, were of existing World War II stock.

On 5 July 1950, early in the conflict, Lieutenant Colonel Charles B. Smith led a task force of US troops to engage a strong North Korean armoured contingent equipped with the formidable Soviet made T-34 tank. 'LTC Smith called on his six supporting howitzers from the 52nd Field Artillery Battalion to pour what artillery men like to call "steel rain" on the enemy,' reported *GI Korea* magazine.

'The howitzers fired their 105mm but the "steel rain" met even stronger iron as the rounds were unable to penetrate the thick armour of the T-34 tanks...

'The tanks continued down the road towards the artillery positions. The artillerymen fired one of their total of nine anti-armour rounds at one of the tanks. There were only nine rounds in the country at the time. The lead tank was hit in the front and burst into flames. The three North Korean tankers

British prime minister Winston Churchill delivers a speech during a visit to the United States. Churchill warned of the advent of the Cold War after World War II ended.

jumped out and fired at an American machine gun position killing the assistant gunner. The assistant gunner would become the first US fatality of the Korean War... The three North Korean tankers were eventually quickly shot down by the other

American. The other tanks were not detoured by the destroyed tank and moved forward. The artillerymen were practically using their howitzers as direct fire weapons at ranges of 150-300 meters at the T34s. One more tank was disabled when it was hit in the treads, but the other tanks kept coming... The artillerymen continued to fire at the tanks as they passed by... They were able to disable another track before all the tanks passed them and continued south.'

During the early years of the Cold War and the Korean Conflict, most of the field artillery employed by the US Army was of World War II vintage, such as the 105mm howitzer M2A1 (also designated the M101), M1 240mm howitzer, M114 155mm howitzer, the 155mm M2 Long Tom and the M115 203mm howitzer. During the Vietnam War, the army introduced the M102 light towed 105mm howitzer, which is also transportable by air and capable of firing a maximum rate of 10 rounds per minute or three rounds per minute sustained at a range of 11,500m (12,576yds). With rocket-assisted ammunition, the weapon can reach more than 15,088m (16,500yds). In 1987, the United States entered a licensing agreement to produce the L118/L119 105mm howitzer of British design, replacing the M102. However, the M102 remains in service with training units of the US Marine Corps and is the standard artillery weapon aboard the famed US Air Force Lockheed AC-130 Pave Spectre gunship.

Other conventional Cold War field artillery in the US arsenal included the M198 155mm gun howitzer, originally deployed in 1979 as a replacement for the M114. The M198 has a maximum range of more than 29,992m (32,800yds) with rocket-assisted ammunition, and

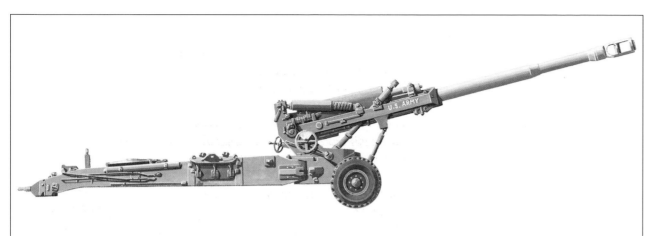

M198 155MM TOWED HOWITZER

Calibre: 155mm (6.1in)
Weight: 7163kg (15.790lb)
Length of piece: 12.3m (40ft 6in)
Elevation: -5° to +72°

Traverse: 45°
Range: 17,800m (19,685yds) with
    standard HE shells; 30,000m (32,810yds)
    with rocket-assisted projectile

devoted to nuclear weapons. In the 1950s, conventional artillery was less of a priority, but in the following decade a resurgence occurred so that as late as the 1960s the majority of Red Army artillery in service remained towed field guns. During the 1970s, the Soviet military and its rivals made a decided turn towards self-propelled weapons.

Soviet towed artillery of the Cold War era included the 122mm gun howitzer D-30, which was deployed in 1967 to replace older weapons of the World War II period. Characterized as simple and reliable, the D-30 could fire a maximum of eight rounds per minute, and one round per minute sustained. Serviced by a crew of eight, its standard range was 15,400m (16,842yds), and that increased to 22,000m (24,059yds) with rocket-assisted ammunition. The 122mm D-74 was an older, heavier 122mm model with a range of just over 23,866m (26,100yds).

17,800m (19,685yds) with standard high-explosive shells. Its crew of nine can achieve a maximum rate of fire of four rounds per minute, and a sustained rate of two rounds per minute. The M198 is also nuclear-capable, evidencing a generation of field artillery designed with the potential for more than conventional warfare in mind.

By the 1990s, it was estimated that at its peak the artillery capacity of the Soviet Red Army included as many as 40,000 guns, howitzers and other weapons. The emphasis on artillery in the Soviet military, although traditional, had waned, the Communist regime depending primarily on the resources

The Vietnam-era M102 105mm howitzer was in service with the US through to Operation Desert Storm. Here artillerymen fire at North Vietnamese positions.

In 1955, the Red Army introduced the nuclear capable 152mm D-20 gun howitzer, which became its standard medium field artillery of the period. The D-20 was also supplied to Egypt, India, North Vietnam and other Soviet client states. Its range with rocket-assisted ammunition was up to 29,992m (32,800yds). The 152mm D-1, a veteran of World War II, remained in service into the 1990s. The 152mm M1976 gun and the slightly modified M1987 have been in service since 1978 as replacements for the 130mm M-46 gun introduced in 1954. The M1976 crew of 11 could reach a maximum rate of fire of eight rounds per minute at a range of up to 24,700m (27,012yds). With its four-wheeled carriage, the M1976 provided protection for its crew with a sloped armour shield. One of its original purposes was to deliver tactical nuclear fire during the close combat conditions sometimes encountered on the battlefield.

The largest towed artillery piece deployed during the Cold War was the Soviet 180mm S-23 gun, which appeared in 1954 and was nuclear capable. Its range with standard ammunition was 30,000m (32,808yds) and with rocket-assisted ammunition up to 43,000m (47,025yds). The Egyptian armed forces employed the S-23 during the 1973 Yom Kippur war with Israel. The M1969 mountain gun is a remaining 76.2mm weapon used in difficult terrain, which can be broken down and transported by individual soldiers.

The Cold War artillery of the British Army included the L118 105mm gun introduced in the 1970s. Commonly referred to as the light gun, it was designed by the Royal Armament Research and Development Establishment and capable of firing six to eight rounds per minute a maximum distance of 17,200m (18,810yds). The L118 replaced various models in service with the British Army during the post-World War II period, including the Italian OTO Melara Mod 56, the 105mm L5 pack howitzer, the elderly 25-pounder and various 75mm guns. A shorter barrel version, designated the L119 is licensed to the United States. Both versions have been in use with the armed forces of numerous countries.

While much of the standard artillery of the British Army during the Cold War included self-propelled models of American design, the 155mm towed FH-70, conceived under a joint venture between Great Britain, the Federal Republic of Germany and, later, Italy, appeared in the 1970s. The Vickers and Rheinmetall firms led the design effort, and the weapon was intended to fire any 155mm ammunition in NATO possession. Its crew of eight can produce a rate of fire of three to six

Troops of the Soviet Red Army fire a Model 1955 D-20 152mm gun-howitzer. The D-20 was one of several heavy Soviet field weapons developed during the Cold War.

rounds per minute or a three-round burst in 15 seconds. Its effective range is up to 30,000m (32,808yds). The artillery arm of the British Army was one of the first to employ digital fire control and spearheaded the development of a cooperative direct fire control system, ABCA, utilized by NATO countries.

Other notable field artillery cannon and howitzers of the West included the 105mm Italian OTO Melara Mod 56, which was developed in the 1950s as a lightweight, mobile howitzer for use by Italian mountain troops. With a maximum range of 11,100m (12,140yds), the howitzer can be disassembled into 12 components and transported by pack animal if necessary. It has been exported to more than 30 countries and remains in active use with a lifespan of more than half a century. The French produced the AUF1 and TR 155mm howitzers during the Cold War to complement various types of US weapons, while West Germany employed the FH-70 155mm howitzer from the late 1970s into the twenty-first century. Gerald Bull's Space Research Corporation of Canada introduced a 155mm howitzer in the 1970s, which came to be known as the GC-45. Highly accurate at short ranges of up to 3000m (3280yds), the GC-45 is capable of placing rounds consistently within a circle of approximately 10m (11yds).

The People's Republic of China produced its own version of the GC-45, the Type 89, and continues to employ a variety of field artillery. For example, the People's Liberation Army introduced a Chinese version of the Soviet D-20 in 1966. During the 1980s, the Type 83 152mm gun and the PLL01 155mm gun howitzer became standard issue. The Type 83 is serviced by a crew of up to 11 and has a range of up to 1170m (1280yds). While

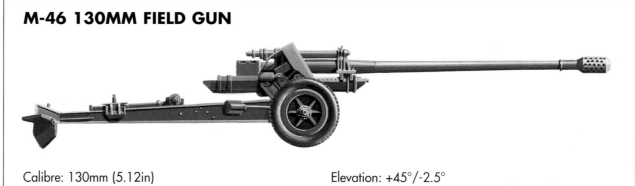

M-46 130MM FIELD GUN

Calibre: 130mm (5.12in)
Weight: travelling: 8450kg (18,629lb);
    firing: 7700kg (16,975lb)
Dimensions: (travelling) length: 11.73m (38.5ft);
    width 2.45m (8.04ft); height 2.55m (8.4ft)

Elevation: +45°/-2.5°
Traverse: 50°
Muzzle velocity: 930 m/sec (3051ft/sec)
Maximum range: 27,150m (29,690yds)

the PLL01, with a crew of five, a semi-automatic loading system, electro-optical sighting for direct or indirect fire control, and a range of up to 38,953m (42,600yds), may be its eventual replacement, large numbers of both are expected to continue in service for some time. The PLL01 rate of fire is two rounds per minute sustained up to a maximum of four.

Although numerous types of artillery ammunition, including high explosive and armour-piercing, had been in service since the beginning of the twentieth century, a new generation of advanced munitions had begun to appear during the 1960s and 1970s. Among these were shells or projectiles that were capable of spreading 'submunitions' over a wide space. These submunitions are often anti-personnel or anti-tank bomblets and mines that open at a prescribed height above the target area and rain their contents on the ground below. Utilizing a forward observer equipped with a laser, targets may be 'painted', or

illuminated, and then struck by tactical guided missiles, providing a degree of standoff capability.

The decline of colonialism

Throughout what came to be known as the Third World, nationalistic fervour and political ideology powered a rise of independence movements across the globe in the decades following World War II. European colonialism was assailed in Southeast Asia, India, the African continent and the Middle East. In the spring of 1954, the communist Viet Minh surrounded the French base at Dien Bien Phu in what was then known as Indochina, and won a victory with far-reaching consequences. Artillery was a major component of the plan for victory conceived by Viet Minh military leader General Vo Nguyen Giap. The Viet Minh massed more than 50,000 troops in the hills surrounding the French positions at Dien Bien Phu and managed to bring 200 pieces of heavy

HEAD TO HEAD: *122mm Soviet D30* VERSUS

The D-30 122mm howitzer entered service with the Soviet Red Army in the 1960s as a replacement for the antiquated M-30 howitzer. Based on a World War II-era German design, the weapon is no longer in production, but remains in service with a number of military forces around the world.

122mm Soviet D30

Calibre: 122mm (4.8in)
Weight in action: 3150kg (6945lb)
Gun length: 40 calibre: 4.875m (16ft)
Elevation: -7° to +70°
Traverse: 360°
Shell type & weight: HE; 21.76kg (47.97lb)
Muzzle velocity: 690m/sec (2264ft/sec)
Maximum range: 15,400m (16,840yds)

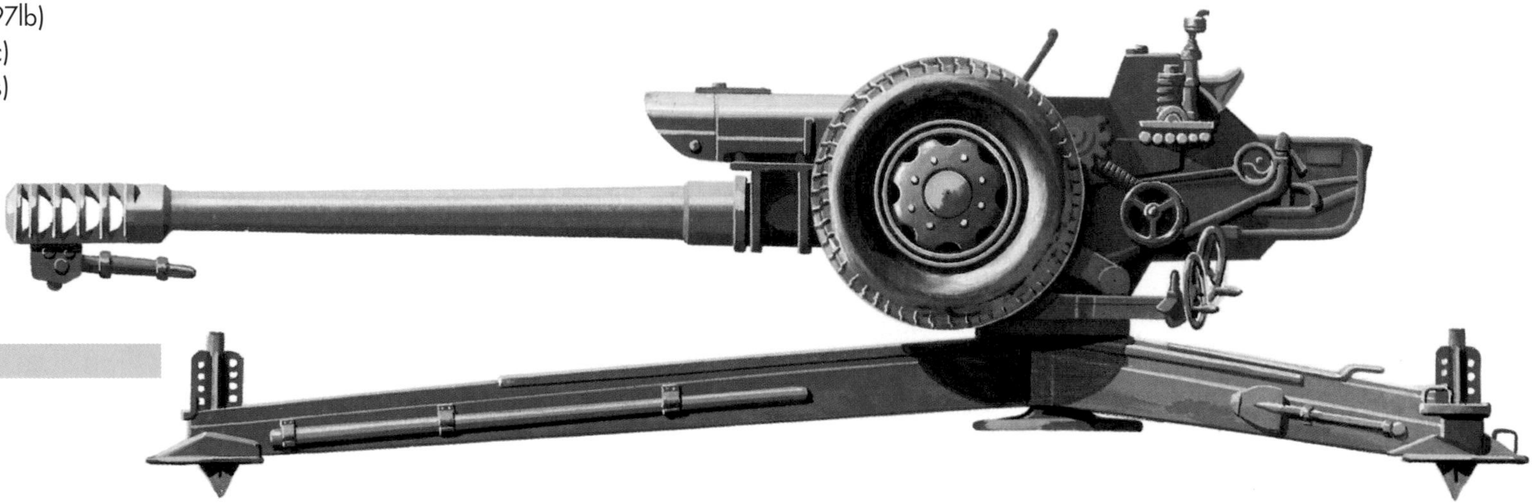

STRENGTHS

- Traverse of 360 degrees
- Anti-tank capability
- Improved range

WEAKNESSES

- Lengthy tactical setup
- Difficult tow
- Low sustained rate of fire

FH-70

The culmination of a cooperative design effort between Great Britain, Germany and Italy, the FH-70, or Field Howitzer for the 1970s, entered service in 1978. Several hundred of the 155mm close support guns were purchased by a number of countries and deployed during the 1980s.

FH-70

Calibre: 155mm (6.1in)
Weight: travelling and firing: 9300kg (20,503lb)
Dimensions (travelling): length: 9.8m (32.1ft);
    width: 2.2m (7.2ft); height: 2.56m (8.4ft)
Elevation: +70°/-3°
Traverse: total 56°
Shell type & weight: HE L15; 43.5kg (96lb)
Muzzle velocity: 827m/sec (2713ft/sec)
Maximum range: 24km (15 miles)

STRENGTHS

- Good rate of fire
- Minimal crew requirement
- Common NATO ammunition

WEAKNESSES

- Heavy weight
- Limited deployment
- Short service life

Chinese Powerhouse

People's Republic of China armaments manufacturers developed the Type 89 155mm field howitzer during the 1980s, basing their design on the CG-45, an innovative artillery weapon that was conceived by Gerald Bull's Space Research Corporation of Canada. The Type 89, shown here during live fire training exercises, has been considered a highly accurate weapon. Bull's reputation as an innovator was somewhat overshadowed by accusations of criminal activity.

artillery within range. After a 55-day siege, characterized by heavy artillery bombardment, the French surrendered.

'Not only were they [the French] outnumbered and outgunned by the enemy artillery,' wrote military analyst Major Harry T. Bloomer, 'but they were also shocked by their inability to destroy enemy artillery.' Colonel Charles Piroth, the French artillery commander, was so distraught over the failure of his own guns to silence the enemy batteries that he committed suicide in his bunker.

Bernard Fall explains further:

'The real surprise to the French was not that the communists had that kind of artillery. What surprised the French completely was the Viet Minh's ability to transport a considerable mass of heavy artillery pieces across roadless mountains to Dien Bien Phu and to keep it supplied with a sufficient amount of ammunition to make the huge effort worthwhile.'

The backbone of the artillery employed by the Viet Minh at Dien Bien Phu, according to researcher Pierre Asselin, was a complement of 24 105mm howitzers of American manufacture, which had been captured by the People's Liberation Army of China during the Korean War. Fourteen years later, the North Vietnamese Army employed similar tactics against the American base at Khe Sanh during the Vietnam War. This time, tactical air power, effective counterbattery fire and a successful relief operation succeeded in raising the siege.

Atomic Annie

Among the military planners of the Warsaw Pact and NATO during the Cold War, the belief prevailed that neither side could employ tactical nuclear weapons without inevitably progressing to

General Vo Nguyen Giap, architect of the Viet Minh victory at Dien Bien Phu, utilized concentrated artillery fire to pound French positions.

1953, Atomic Annie – a specially configured 280mm gun that had been constructed at Fort Sill, Oklahoma, by the army's Artillery Test Unit – fired a28cm (11in) shell armed with a fission warhead with an estimated payload of 15 kilotons. Atomic Annie, designated a T-131 cannon, was situated on the T-10 heavy artillery transporter (M65), a bridgelike structure carried above a pair of 6 by 6 tractors. The weapon is on display today at the US Artillery Museum at Fort Sill. The shell travelled 10,000m (10,936yds) and detonated 160m (525ft) above ground at the Nevada Test Site. Witnessed by about 3200 observers, the test, codenamed Shot Grable and conducted as part of Operation Upshot Knothole, was a success and remains the only atomic artillery shot fired during the US testing programme.

the launch of larger nuclear weapons and eventually intercontinental ballistic missiles (ICBM). Research and development of tactical nuclear weapons began in earnest during the 1950s, continued throughout the Cold War, and resulted in the development of nuclear guns, rockets, missiles and even land mines.

The US Army adapted guns and howitzers to deliver tactical nuclear artillery shells. On 25 May

Viet Minh troops charge towards French positions at Dien Bien Phu. The 1954 Communist victory effectively ended French colonialism in Southeast Asia.

From 1950 to 1990, a series of tactical nuclear warheads were produced by the United States. The warhead fired during Shot Grable was designated W9 and was one of 80 produced in 1952–53 for the conventional T-124 shell. The W9 was replaced in 1957 by the W19, a longer version of the W9. Eighty of these were also produced and then retired in 1963. This was followed by the W48, which weighed nearly 58kg (128lb) and was fired from a 155mm howitzer. The nuclear warhead was affixed to the Artillery Fired Atomic Projectile (AFAP) M-45. Its yield was the relatively low equal of 65 tonnes (72 tons) of TNT. During the 1960s, more than 1000 of the W48 and its Mod 0 and Mod 1 variants were produced and eventually deactivated. From 1957 to 1965, approximately 2000 W33 warheads were produced for use with 203mm shells. The W33 nuclear-armed shell weighed just over 108kg (240lb). Plans to develop the W74, with a 91-tonne (100-ton) explosive yield, for the 203mm shell were cancelled in 1973. In the 1980s, the W82 and the fission type W82-1 were developed to arm the XM-785 155mm shell, but both projects were ended by 1990.

The Soviet Union developed numerous warheads and weapons systems to deliver tactical nuclear weapons under the control of the 12th Main Directorate of the Ministry of Defence. These weapons were deployed with the Red Army in battlefield units at division level.

Early in the 1990s, the United States and Russia agreed to remove their nuclear-armed artillery

Shown atop its heavy wheel prime mover, the US 280mm nuclear capable cannon nicknamed 'Atomic Annie' is pictured.

Tactical Nuclear Threat

During the 1950s, both the NATO countries and those of the Warsaw Pact determined that tactical nuclear weapons, which could either destroy an opposing force or render it incapable of further resistance, were a potentially viable component of defensive planning. For the United States and its allies, it was apparent that a major Soviet ground offensive in Europe could be disastrous in the short term due to the sheer overwhelming numerical superiority of the Red Army. Eventually, a series of negotiated treaties resulted in the removal of most or all of the tactical nuclear weapons deployed in Europe, although recent evidence indicates that the Soviets failed to comply with many of the terms agreed to in these treaties.

shells from active duty. More than 1300 were taken out of service in Europe by the United States alone.

Shoot and scoot

During World War II, the importance of mobile artillery had been established as both the Allies and the Axis used self-propelled weapons for direct fire support. Not only was it recognized that self-propelled artillery could provide firepower in a timely manner, but also that its mobility was a key element in protecting it from counter-battery fire. As the shift towards mobile artillery gained momentum, systems were configured with proven guns married to proven tank and armoured vehicle chassis or by overall integral design from top to bottom. An array of weaponry was deployed, including guns, howitzers, rockets and missiles, which were capable of firing both conventional and nuclear weapons.

Through the decades, refinements have occurred in self-propelled artillery, producing even greater speed and mobility, increased accuracy and more robust firepower. Modern armies have come to rely on their self-propelled artillery to play a decisive role on the battlefield.

In conjunction with the proliferation of its successful M7 Priest design, the United States introduced several self-propelled artillery systems during the 30 years after the end of World War II. The M40 GMC (Gun Motor Carriage) was developed during World War II as a replacement for the M12, of which only about 100 were built from 1942 to 1943. The M40 participated in limited action near the end of the war, but was most widely deployed during the Korean War five years later. The vehicle mounted a 155mm M2 gun on the chassis of the M4A3 medium tank. It weighed 33 tonnes (36.3 tons), was serviced by a crew of eight, and was capable of reaching 40km/h (23.6mph) on the road or 30km/h (14.29mph) offroad powered by a 340hp Wright Continental R975 EC2 engine.

In the early 1960s, the United States produced both the M108 and M109 self-propelled howitzers. The M108 was armed with an M103 105mm howitzer and serviced by a crew of five. The vehicle weighed 19 tonnes (21 tons) and was capable of reaching a top speed of 56km/h (35mph) powered by a 405hp Detroit diesel turbocharged 8V-71T engine. Secondary armament consisted of a single M2 0.5-calibre machine gun mounted in the turret. Incorporating the same turret as that of the M109 and components of the chassis of the M113 armoured personnel carrier, the M108 was placed in service with numerous NATO countries. Early in the Vietnam War, however, the decision was made to phase the M108 out of active duty with the US Army in favour of the heavier gunned M109. The standard main armament of the M109 is a 155mm howitzer.

The M107 175mm self-propelled gun was also introduced during the 1960s, and the majority of its combat service took place during the Vietnam War. The vehicle consisted of the gun mounted atop an open carriage, which provided little cover for the crew of 13 but did contribute to a favourable weight of 25.6 tonnes (28.3 tons). The M107 was capable of firing a 67kg (147.7lb) projectile a distance of 30,000m (32,808yds) at a top speed of 56km/h (34.8mph). The M107, therefore, provided excellent fire support and mobility but sacrificed armour for speed. A number of M107s were used by the Israeli Defence Force during its conflicts with neighbouring Arab states. Although the vehicle is still in use by numerous countries, it was withdrawn from service with the US Army prior to 1980.

In 1963, the M110 203mm howitzer was deployed by the US Army. During its length of service, the vehicle participated in the Vietnam War, Operation Desert Shield and Operation Desert Storm. It has served with a number of other nations, including Great Britain, Japan, Greece, Spain and Turkey. Serviced by a crew of 13, the M110 is capable of firing a rocket-assisted round up to 30km (18.64 miles). Its maximum speed is 55km/h (34.18mph), and during its tenure the M110 was the largest self-propelled howitzer in use with the US Army. Mounted on a chassis virtually identical to that of the M107, the weapon could fire a maximum of three rounds per minute and one round every two minutes sustained. Interestingly, the barrels of the large howitzer have

been put to use in making casings for the GBU-28 bunker buster bomb.

Two other Cold War self-propelled artillery pieces are worthy of mention. The US M44 155mm gun and the M52 105mm howitzer have both been retired. Development of the 28-tonne (31-ton) M44 prototype began as early as 1952, and production began the following year. Capable of reaching a top speed of 56km/h (35mph) on the road or 18km/h (11 miles) per hour cross-country, the vehicle may fire rocket-assisted shells up to 30,000m (32,808yds). Its secondary armament consists of a single .50-calibre machine gun. Although it remains in service with

numerous countries, the M44 was withdrawn from active duty with the US Army in the early 1960s in favour of the M109. During the 1980s, a joint venture between the German companies Rheinmetall, GLS and MTU upgraded the M44 with an improved gun and a diesel engine, resulting in the M52. Weighing just over 24,494kg (54,000lb), the M52 was placed on active duty in 1955 after several years of research. Serviced by a crew of five, it used the same chassis as the M44 and the M41 Walker Bulldog light tank. Its Continental AOS-895-3 engine provided 500hp, and its secondary armament consisted of a .50-calibre machine gun. Nearly

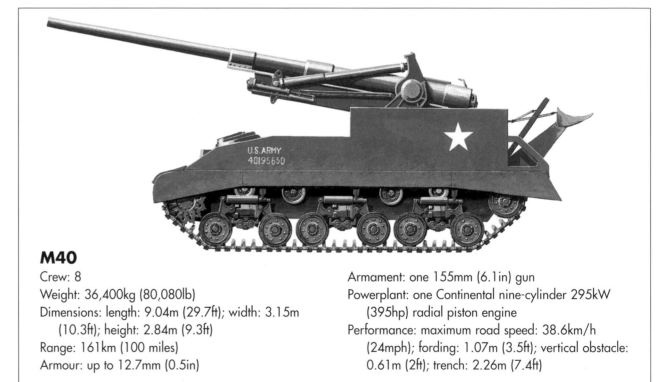

M40

Crew: 8
Weight: 36,400kg (80,080lb)
Dimensions: length: 9.04m (29.7ft); width: 3.15m
    (10.3ft); height: 2.84m (9.3ft)
Range: 161km (100 miles)
Armour: up to 12.7mm (0.5in)

Armament: one 155mm (6.1in) gun
Powerplant: one Continental nine-cylinder 295kW
    (395hp) radial piston engine
Performance: maximum road speed: 38.6km/h
    (24mph); fording: 1.07m (3.5ft); vertical obstacle:
    0.61m (2ft); trench: 2.26m (7.4ft)

Above: The high silhouette of the US-made M52 105mm self-propelled howitzer is apparent in this rear view showing the enclosed crew compartment.

Right: The American M44 155mm self-propelled was introduced in the 1950s. Now withdrawn from the US Army, it still serves with other countries.

700 of the model were built for the US Army, and it was widely used with NATO forces and during the Vietnam War.

Soviet mobility

Variants of Soviet self-propelled artillery deployed in World War II continued to serve with the Red Army and other forces of the Warsaw Pact well into the decades of the Cold War. In some cases, self-propelled artillery completely replaced certain models of towed artillery. Several more recent systems are representative of the Soviet and Warsaw Pact self-propelled artillery of the time.

During the 1960s, the Soviets developed the 2S1 Gvozdika (Carnation), a 122mm self-propelled howitzer. The 2S1 was placed in service in the early 1970s and first identified by the West during a parade conducted by the Polish Army in 1974.

Powered by a YaMZ-238N diesel engine generating 300hp, the vehicle is capable of speeds up to 160km/h (37.28mph) on the road and 30km/h (18.64mph) cross-country. It has been produced in the Soviet Union, Poland and Bulgaria. Several variants, including vehicles for mine-clearing, air-defence management, electronic jamming, and command, have also been produced.

The 2S3 Akatsiya (Acacia) was developed in the late 1960s and designated by NATO as the M1973. With a main armament of a 152mm gun and secondary armament consisting of one 7.62mm machine gun, the vehicle weighs 27,500kg (60,627lb) and is capable of a top speed of 60km/h (37.28mph) on the road powered by a V-59 V-12 diesel engine. Protected by armour up to 2cm (0.79in) thick, the system is serviced by a crew of six. The gun is capable of firing a high explosive shell a distance of 18,500m (20,232yds) or a rocket-assisted projectile up to 24,000m (26,246yds). Several variants have remained in service with the armed forces of the former Warsaw Pact and other nations, such as Libya, Syria, Iraq and Cuba, since the 1970s.

Designated the M1975 by NATO, the 2S4 Tyulpan (Tulip) is a 240mm self-propelled mortar that is the heaviest in the arsenal of any active military force today. Capable of reaching a top speed of 262km/h (38.53mph) on the road, the Tyulpan is powered by a 520hp V-59 diesel engine. Secondary armament consists of one 12.7mm DShK machine gun, and the vehicle weighs 27 tonnes (30 tons). In transit, the weapon is operated by a crew of four; however, an additional five men are needed to service the mortar during a firing sequence. The mortar itself is attached to the hull of the GMZ tracked mine-laying vehicle and fires a projectile weighing 130kg (286.6lb) at a rate of one

round per minute. With conventional projectiles, its range is 9650m (10,553yds), and rocket-assisted shells extend that to 20,000m (21,872yds). The Tulip also fires armour-piercing, nuclear and laser-guided rounds.

The 2S5 Giatsint (Hyacinth) became operational in 1981 and is known to NATO as the M1981. This powerful self-propelled vehicle mounts a 152mm gun with a single 7.62mm machine gun as secondary armament. Powered by a

520hp 388 kW diesel engine, the Hyacinth is capable of speeds up to 62km/h (38.53mph) on the road and 25km/h (15.53mph) cross country. The vehicle weighs 25.5 tonnes (28.2 tons) and is serviced by a crew of five. Nearly 600 of the weapons have been produced, and they have been placed in service by nations of the former Soviet Bloc and exported to Finland.

During the 1970s, the DANA ShKH77 was designed and built by engineers in the former

The Soviet self-propelled 2S5 Giatsint with a 152mm cannon, also known to NATO as the M1981, was developed during the early 1970s.

Warsaw Pact nation of Czechoslovakia. First appearing in the early 1980s, the DANA was innovative in its wheeled rather than tracked mobility. The vehicle incorporates a 152mm gun atop the platform of the Tatra 815 truck. With a crew of five, the DANA was conceived to provide indirect fire support to infantry, and offered a cost saving compared to the purchase of the Soviet-made 2S3 Akatsiya self-propelled weapon. The wheeled vehicle proved to be easier to maintain and flexible in mobility due to the fact that tyre pressure can be regulated for optimum performance on varied terrain. Hydraulic stabilizers may be lowered and retracted to assist in firing from a stable platform.

An upgraded variant, the ShKH Zuzana, incorporates a 155mm gun mounted on the chassis of the Soviet T-72 tank. The Slovak Army accepted the Zuzana in 1998, and it conforms with NATO armaments standards. By the mid-1990s, more than 750 of the DANA vehicle had been produced and placed in service with countries such as Poland, Slovakia, Russia, Libya and the Czech Republic.

Three decades of service

In 1964, the FV433 Abbot self-propelled 105mm gun was placed in service with the British Army. Manufactured by Vickers, the Abbot weighed 14.5 tonnes (16.26 tons) and was powered by a 240hp Rolls Royce RK60 6-cylinder engine. With a crew of six, the Abbot entered service about the same time that the heavier American M109 series was deployed. One of a number of variants utilizing the FV430 chassis, the Abbot was in service for three decades before retirement in favour of the AS-90 155mm self-propelled gun in 1995. Secondary armament consisted of a 7.62mm L4A4 machine gun with 1200 rounds of ammunition and smoke

Mounting a 152mm cannon, the wheeled DANA was designed and built in Czechoslovakia during the 1970s as a close infantry support weapon.

dischargers for concealment. It was capable of a top speed of 47km/h (29.2mph). A flotation system attached to the hull provided a degree of amphibious quality to the Abbot under battlefield conditions. A few Abbots were exported to the Indian Army.

Prominent among the self-propelled artillery of the People's Republic of China during the Cold War era is the Type 83 152mm gun howitzer, initially produced in the early 1980s by what was then known as the 674 Factory and is now called the Harbin First Machinery Building Group, Ltd. The Type 83 used by the People's Liberation Army was based on the Type 321 utility chassis and the Type 66 towed artillery gun howitzer. The weapon is also reported to be capable of firing laser-guided

HEAD TO HEAD: *M109* VERSUS

The basic design of the M109 self-propelled howitzer has served as the backbone of the US Army's mobile field artillery for more than three decades. With its 155mm weapon, the system is powerful and adaptive to varied climates and terrain. The current version, the M109A6 Paladin, is used by the US only. Other nations have begun to field German and British designs.

M109

Crew: 6
Weight: 23,723kg (52,192lb)
Dimensions: length: 6.61m (21.7ft); width: 3.3m
    (10.83ft); height: 3.29m (10.79ft)
Range: 390km (240miles)
Armour: classified
Armament: one 155mm (6.1in) howitzer; one
    12.7mm (0.5in) anti-aircraft machine gun
Powerplant: one Detroit diesel 302kW (405hp) engine
Performance: maximum road speed: 56km/h
    (35mph); fording: 1.07m (3.5ft); vertical obstacle:
    0.53m (1.7ft); trench: 1.83m (6ft)

STRENGTHS

• Service longevity
• Excellent firepower
• Manoeuvrability

WEAKNESSES

• High cost
• Technological upgrades needed
• Wear due to attrition

Soviet 2S3

Introduced into the Soviet military in 1973, the 2S3 M1973 Akatsiya, known in NATO parlance as the Acacia, combines a modified version of the 152mm D20 gun howitzer and a chassis similar to the SA-4/GANEF. The weapon is similar in appearance to the US M109.

Soviet 2S3

Crew: 6
Weight: 24,945kg (54,880lb)
Dimensions: length: 8.4m (27.6ft); width: 3.2m (10.5ft); height: 2.8m (9.2ft)
Range: 300km (186 miles)
Armour: 15–20mm (0.59–0.78in)
Armament: one 152mm (6in) gun; one 7.62mm (0.3in) anti-aircraft machine gun
Powerplant: one V-12 388kW (520hp) diesel engine
Performance: maximum road speed: 55hm/h (34mph); fording: 1.5m (4.9ft); vertical obstacle: 1.1m (3.6ft); trench 2.5m (8.2ft)

STRENGTHS

- Highly mobile
- All-terrain effectiveness
- Good range

WEAKNESSES

- Restricted elevation
- Thin armour
- Non-amphibious

Introduced in the 1960s, the British Abbot 105mm self-propelled gun was in service for more than 30 years prior to being phased out by the AS-90.

munitions such as the 152mm rockets produced in China. Along with the 674 Factory, several other production facilities in the People's Republic were responsible for components of the weapon, whose prototype was tested in 1980.

From 1984 to 1990, nearly 80 of the Type 83 were produced to provide direct fire support to armoured troops and motorized infantry. Powered by a 387.8kW (520hp) WR4B-12V150LB diesel engine, the Type 83 is capable of a top speed of 55km/h (34.18mph). The maximum rate of fire for the 152mm gun howitzer is five rounds per minute with conventional projectiles. Secondary armament consists of a single 12.7mm machine gun mounted on the turret and a 40mm rocket-propelled grenade launcher. First identified by the West in 1984, the Type 83 is assumed to remain in service today.

The principal Cold War self-propelled artillery of the French Army consisted primarily of variants of the AMX-13 light tank, which was in service from 1953 to 1985. During the Cold War era, more than 100 variants of the AMX-13 were produced and more than 7700 individual units were in service with the armed forces of France and many other countries. The tank itself was withdrawn from service in the mid-1970s, but the self-propelled artillery variant in 105mm and 155mm configurations continues to be active. The weight of the weapon is 13 tonnes (14.5 tons), and it is powered by a SOFAM Model 8 Gxb 184.4kW (250hp) engine capable of a top speed of 60km/h (37.28mph). Secondary armament consists of a 7.5mm or 7.62mm machine gun and smoke grenade launcher.

The Israeli Defence Force employed a number of self-propelled artillery designs during the Cold War period. Among the principal weapons are the L-33 Ro'em, consisting of a 155mm howitzer atop the chassis of the M4 Sherman tank. Produced by the Israeli manufacturing firm of Soltan, the vehicle debuted in 1968 and served during the Yom Kippur War of 1973. Weighing 37.6 tonnes (41.5 tons), the L-33 was serviced by a crew of eight. Secondary armament included one 7.62mm machine gun.

The Makmat 160, a self-propelled 160mm mortar, was developed following the Six-Day War of 1967 and served during the Yom Kippur War as well. Entering service in 1968, the weapon also incorporated the M4 Sherman tank chassis, weighed 32.6 tonnes (36 tons), and was serviced by a crew of up to eight. Between 1984 and 1986, Soltam built a 155mm self-propelled howitzer prototype. Known as the Sholef, or Gunslinger, the 41 tonne (45 ton) vehicle was never placed in production, and the Israelis continued to purchase the American-built M109 series.

Recoilless guns

After World War II, the generation of heavy anti-tank guns developed during that conflict remained the standard for years. Many of these were adapted to a self-propelled role and mounted on mobile tank chassis or wheels for added mobility. Although little changed in the configuration of towed anti-tank weapons themselves, the projectiles fired by them continued to evolve. Shells with a tungsten core, which could penetrate thick armour at high velocity; the discarding sabot projectile, which enabled the tungsten warhead to achieve that velocity; and the

advent of the hollow or shaped charge continued in widespread use.

Two developments in anti-tank weaponry of the Cold War period worthy of note were the proliferation of recoilless guns for use against armour and the development of tank-killing wire or laser-guided missiles that could be fired from a jeep or tracked mount, or even from the shoulder of an infantrymen, harkening back to the days of the bazooka and the German panzerfaust of World War II. Although the recoilless weapon was nothing new, having been successfully test-fired well before World War I, its appearance on the battlefield did not take place until World War II when such

weapons up to 105mm were deployed. The recoilless principle by that time was practically achieved by an apparatus that dispersed as much as 80 per cent of the gases generated by the ignition of the projectile backwards through a venting system. These weapons became standard among the armies of the major powers during the post-war years despite the fact that they were difficult to fire and manoeuvre due to the large backblast, or signature, of the discharged weapon.

As French military involvement in Vietnam and Algeria escalated during the late 1940s, the US 75mm recoilless rifle of World War II vintage was used as an anti-tank weapon by the French Army,

AMX-13 DCA

Crew: 3
Weight: 17,200kg (37,840lb)
Dimensions: length: 5.4m (17.7ft); width: 2.5m (8.2ft); height (radar up): 3.8m (12.5ft); height (radar down): 3m (9.8ft)
Range: 300km (186 miles)
Armour: 25mm (0.98in)
Armament: twin 30mm (1.18in) Hispano cannon
Powerplant: one SOFAM eight-cylinder water cooled 186kW (250hp) petrol engine
Performance: maximum road speed: 60km/h (37mph); fording: 0.6m (1.97ft); vertical obstacle: 0.65m (2.1ft); trench: 1.7m (5.56ft)

which placed the gun on a Vespa scooter. The US Army made extensive use of the 57mm, 75mm and 105mm recoilless rifles during the Korean War, while the armies of Great Britain and the Soviet Union deployed recoilless rifles in significant numbers as well.

Perhaps the most widely known recoilless rifle is the Swedish manufactured Carl Gustav, which is still in use around the world today. A product of the legendary Bofors company and the designers Harald Jentzen and Hugo Abramson of the Royal Swedish Arms Administration, development of the Carl Gustav began shortly after World War II. By 1948, it was in service with the Swedish military. With a rifled barrel, which stabilized the projectile, it was considerably more powerful than its counterparts in service with other armies. The 84mm shoulder-fired weapon was exported in large numbers, and during the 1960s an improved variant, the M2, was introduced. The M3, which entered service in 1991, replaced many wearing parts with modern lightweight materials.

The special forces units of numerous countries still use the Carl Gustav in action against armour or hardened bunkers. It can be operated by a single soldier or a pair of soldiers working as a team, and fires at a rate of up to six rounds per minute. Telescopic sights, laser range-finding, and image intensification systems maintain a high level of accuracy. The Carl Gustav has seen action as recently as NATO operations in Afghanistan.

The Soviet Red Army continued its use of hybrid artillery intended for infantry support or for the anti-tank role during the Cold War. However, enthusiasm for the recoilless rifle is apparent in the number of such weapons placed in service. Replacing the SPG-82, the B-10 recoilless rifle

Sherman's Descendant

The ubiquitous M4 Sherman tank, a World War II vintage weapon, survived in service for decades with various revisions. Its chassis provided the mobility for the Soltan L-33 Ro'em, a 155mm self-propelled howitzer that was designed in Israel and entered service with the Israeli Defence Force in 1968. The L-33 was deployed during the Yom Kippur War of 1973. One of the design's drawbacks was its high silhouette, which presented a large target for enemy gunners.

became operational in 1954. Its crew of four fired an 82mm (3.2in) projectile at a rate of up to seven rounds per minute, and the B-10 was exported to Soviet Bloc nations, as well as Egypt, North Korea, Pakistan and Syria. The North Vietnamese Army and Viet Cong used the B-10 during the Vietnam War. Mounted on a two-wheeled carriage, the B-10 was generally retired from regular service during the 1960s and replaced by the SPG-9, although it is still in use today with airborne units of the Red Army.

The upgunned 2A19, also known as the T-12, was placed in Red Army service in 1955 and fired a 100mm (3.9in) projectile at a rate of up to 14 rounds per minute under stressed combat situations and generally up to 10 rounds per minute in the field. The 2A19 replaced the D-10 model, which had been used in World War II and was usually distributed to infantry units in rifle divisions or attached to armoured regiments. Updated variants such as the 2A29/MT-12 Rapira, which was deployed in 1970, are capable of firing tungsten core, high-explosive fragmentation and laser-guided munitions. The 2A29 was eventually replaced by the heavier 125mm (4.in) 2A45 Sprut B, another smoothbore weapon, due to the need to penetrate ever-increasing thicknesses of armour and stronger steel alloys.

The SPG-9 is a 73mm recoilless gun, which entered service with the Red Army in 1962. Serviced by a two-man team, the weapon could be brought into firing position within a minute and deliver a variety of ordnance at a rate of six rounds per minute with an effective range of up to nearly 800m (875yds) and a maximum range of 6500m (7108yds). Used by the armies of Romania, Bulgaria and Iran, the weapon remains in widespread service.

During the late 1980s, the 2A45 Sprut was produced by the Petrov Design Bureau. This 125mm semi-automatic weapon gave Red Army troops the advantage of mobility over a short distance with the employment of an APU. Conversion from firing position to mobile can be accomplished in about two minutes, and deployment from mobile to firing position in as little as 1.5 minutes, offering a measure of protection against retaliatory fire. Serviced by a crew of seven, the Sprut is capable of delivering up to eight rounds per minute at an effective distance of 12,200m (13,342yds) with standard high-explosive ammunition. An added battlefield

During training, soldiers fire the Carl Gustav, a Swedish anti-tank weapon that may be classified as a recoilless rifle. Developed in the 1940s, it remains in service today.

convenience is that the Sprut fires the same ammunition types as the T-64, T-72, T-80 or T-90 main battle tanks. It is also capable of firing laser-guided ordnance. Both self-propelled and towed variants have been deployed in the armed forces of numerous nations.

Denis Burney

The foremost proponent of recoilless weaponry in Cold War era Great Britain was Denis Burney, who designed the Wallbuster HESH round for use against the Atlantic Wall defences of the Germans in World War II. Although the round was not used during Operation Overlord, Burney produced a number of recoilless weapons, such as an 88mm weapon that could be carried by a single soldier. The Burney-inspired L6 Wombat was the pinnacle of British recoilless weapon development. A

system. Its effective range is 580m (634yds), while its maximum range is nearly 7700m (8421yds). Primary ordnance includes high-explosive and high-explosive anti-tank (HEAT) rounds.

Chinese persistence

The Chinese have never completely abandoned the concept of towed anti-tank weapons and were developing these nearly 40 years after the end of World War II. The Type 73 100mm gun was deployed in 1973 with the People's Liberation Army and was the first Chinese-made gun capable of firing armour-piercing, fin-stabilized, discarding sabot rounds. By 1983, the Type 73 had replaced the obsolete Type 56 85mm gun. In the Soviet style, the Type 73 is intended for both anti-tank and fire-support roles. Its crew of nine can fire at a rate of eight to 10 rounds per minute at an effective range of 1730m (1892yds) and a maximum range of 13,708m (14,982yds).

Another Chinese towed anti-tank gun, the Type 86 100mm weapon, reached People's Liberation Army units in the late 1980s. NORINCO, the state-run controlling agency of arms production and exporting, claimed that its armour-piercing round was capable of compromising the frontal armour of the T-72 tank at 2000m (2187yds). Also crewed by nine soldiers, the Type 86 fires up to 10 rounds per minute at a maximum range of nearly 15,000 yards. Considered a primary anti-tank weapon, the gun is deployed with motorized infantry formations of the People's Liberation Army.

120mm rifle, the Wombat, replaced the BAT and MoBAT, which were earlier 88mm weapons. Eventually supplanted by anti-tank missiles, the Wombat was mounted on a two-wheeled carriage and towed by various vehicles or mounted as a self-propelled weapon aboard the FV432 chassis. With an effective range of slightly less than 1005m (1100yds), the Wombat could fire on targets up to twice that distance with the assistance of a forward observer.

During the 1970s, the People's Republic of China equipped the People's Liberation Army with the Type 75 105mm recoilless gun mounted atop the four-wheel drive Beijing BJ2020S vehicle. Entering service in 1975, the Type 75 has been used by both infantry and airborne units and can be airlifted by helicopter to combat zones. Crewed by up to five soldiers, the weapon can fire up to six rounds per minute utilizing optical sights, laser range-finding, and a computer-based fire control

Representative of the recoilless rifles in use with the US Army during the Cold War are the various models of 57mm, 75mm and 105mm weapons. The M20 75mm recoilless rifle weighed 52kg (114.6lb) and was mounted on the tripod of a machine gun. Capable of destroying a T-34 medium tank at a distance of 366m (400yds), it was much more effective against armour than its 57mm predecessor, which could also be shoulder-fired. The maximum range of the M20 was 6400m (7000yds).

The M40 recoilless rifle was commonly mounted on an M151 Jeep and provided mobile anti-tank artillery protection for smaller US Army formations. The 105mm weapon was commonly referred to as a 106mm gun in a deliberate attempt to avoid confusion with the ammunition of the cancelled M27 system, which was incompatible with the M40. The gun has been mounted on M274 Mechanical Mules, Toyota Land Cruisers, Land Rovers, the M113 chassis and the HMMWV vehicle. Prior to being replaced by the BGM-71 TOW anti-tank missile, the M40 was used extensively during the Vietnam War. A .50-calibre spotting rifle was attached to the M40 and used a specially developed round for simulating the flight path of the larger shell. The M40 is still in use with the armies of Turkey, Taiwan, Israel, Mexico, Morocco, South Korea and other countries.

A self-propelled variant of the M40, the M50 Ontos, mounted six of the 105mm recoilless rifles atop a tank chassis. The Ontos provided outstanding firepower, especially in the close-quarters combat during the Vietnam War at places such as Hue. However, its rifles had to be serviced from an exposed position, outside the relative safety of the hull, and its tremendous backblast drew the

Replaced by anti-tank guided missiles, the 120mm L-6 Wombat recoilless rifle was often mounted on a two-wheeled carriage.

attention of enemy batteries and anti-tank weapons. The Ontos was effective against ensconced enemy snipers and in clearing buildings of hostile troops. Introduced in the late 1950s, it was used throughout the Vietnam War and as late as the US intervention in Lebanon in the early 1980s.

The recoilless rifle entered the nuclear age in the late 1950s when the US Mk-54 Davy Crockett, a small fission warhead, was developed to be fired by a recoilless rifle. The bulbous warhead weighed about 23kg (51lb) and was 79cm (31in) long with a diameter of 28cm (11in). It was first test-fired during Operation Hardtack in October 1958. Test firing occurred again during the Little Feller shots of July 1962. Its initial explosive yield was 9 tonnes (10 tons), but this was later enhanced to a full

The US M50 Ontos was developed in the 1950s and saw service in Vietnam. Mounting six M40 106mm recoilless rifles, it was capable of delivering tremendous fire.

kiloton. About 400 Davy Crockett weapons were produced in the early 1960s, but by 1971 these had virtually been withdrawn from frontline units of the US Army.

Theoretically, the Davy Crockett was designed for use with NATO forces in West Germany. In the event of a Soviet attack, the weapons could be used to blunt the enemy offensive and contaminate the area surrounding the impact of the projectile for at least 48 hours, buying time for NATO forces to mobilize. The Davy Crockett projectile could also be fired from 120mm or 155mm launchers. A total of 2100 were eventually built.

Guided weapons

Anti-tank guided missiles (ATGM) or anti-tank guided weapons (ATGW) are in continuous development around the world. Many of these may either be shoulder-launched by individuals or teams of soldiers, operated from self-propelled armoured vehicles, or mounted on lightweight vehicles like the Jeep. These weapons offer infantry the opportunity to take out enemy armour with a single shot at much greater range than their World War II predecessors such as the US bazooka, the British PIAT, or the German panzerfaust. A great battlefield equalizer, the ATGM or ATGW allows the infantryman to fire and maintain concealment or fire and vacate his position.

Many of these weapons are guided by laser or by wire, evolving from the early X-4 and X-7 models produced in Nazi Germany. Manually command guided missiles (MCLOS) have been largely

superseded by a second generation of weapons designated SACLOS, or semi-automatically command guided systems, and eventually the still more complex and most recent 'fire and forget' systems of today. Successive improvements have allowed infantry personnel to become less and less

vulnerable. The MCLOS requires the operator to maintain position and use a joystick to guide the missile, while the SACLOS requires the operator to maintain the target in his sights, and the 'fire and forget' allows totally free movement after launching.

During the Cold War, the British developed the

One of two recoilless rifle systems capable of delivering the 1950s M388 Davy Crockett tactical nuclear projectile, the 4in M28 was deployed in Western Europe.

Malkara and Vigilant missiles during the late 1950s. The Vickers Vigilant fired a 14kg (31lb) HEAT warhead contained in a 1m (3.5ft) missile with a solid fuel propulsion system. Its range was up to 1375m (1504yds), and the weapon was guided by a wire of about 63m (69yds) in length. Mounted on the Ferret armoured car or the Land Rover, the Vigilant was supplanted by the Swingfire in the 1970s.

During the 1950s, France and Switzerland introduced their Entac and Cobra missiles, prompting the Soviet Union to accelerate its programmes. The result was an improvement over the existing AT-1 Snapper, which NATO called the AT-3 Sagger. Officially designated as the 9M14 Malyutka (Little Baby), the Sagger was widely produced and exported during the 1960s and 1970s, and more than 25,000 were reported to have been manufactured on an annual basis.

Development of the MCLOS wire-guided Sagger began in 1961, and the missile was placed in service by the end of 1963. Delivering a warhead of just over 10kg (22lb) housed in a missile of nearly 0.9m (3ft) in length, the effective range of the weapon is up to 3000m (3280yds). Beginning in 1972, the Sagger saw action with the North Vietnamese Army during the Vietnam War, and it was responsible for the destruction of a large number of Israeli tanks during the 1973 Yom Kippur War while in service with the Egyptian and Syrian armies. The missile could be deployed within five minutes from a launcher the size of a suitcase, mounted on a vehicle, or fired from a helicopter. Later versions were improved with a SACLOS guidance system.

Another Soviet anti-tank missile of the Cold War was the AT-7 Saxhorn, or 9K115 Metis (Mongrel) as it was called by the Red Army. The Saxhorn was lighter and less complicated to operate than its predecessor, the AT-4 Spigot system. Delivering a HEAT warhead of 2.5kg (5.5lb) a maximum range of nearly 1005m (1100yds), the Saxhorn is 0.74m (2.4ft) in length and weighs 5.5kg (12.1lb). Introduced to Red Army service in 1979, the weapon is operated by a two-man team.

Designed in the 1950s, the Vickers Vigilant was a wire-guided anti-tank missile that could be delivered by infantry or mounted atop light vehicles.

From 1963 to 1968, the Hughes Aircraft Company worked to develop the prototype of the BGM-71 TOW (Tube launched, Optically tracked, Wire command link guided) missile. The weapon was approved in the late 1960s and placed in service with the US military in 1970 to replace the M40 recoilless rifle and the antiquated MGM-32 ENTAC missile system. The TOW was used extensively during the Vietnam War and is capable of being fired from numerous launch platforms, including the M220 launcher mounted on the M151 Jeep, the M113 armoured personnel carrier, the M966 HMMWV vehicle, the Bradley fighting vehicle, the M1134 Stryker ATGM carrier, the AH-1 Apache helicopter – and an individual soldier if absolutely necessary.

In the past 30 years, several improvements have been made to the TOW, including an updated variant in 1978, the TOW 2 in 1983 and the TOW 2A/B in 1987. A SACLOS weapon, the TOW was also upgraded as recently as 2001 and used as recently as Operation Iraqi Freedom. The TOW is

167

1.2m (3.9ft) long, weighs 26kg (57.3lb) and has a range of 3749m (4100yds). Recently, Raytheon has taken over production of the TOW, and it remains the most widely employed anti-tank missile in the world, serving with the armies of no fewer than 45 countries.

In 1975, the US Army deployed the M47 Dragon shoulder-fired anti-tank missile, which was manufactured by McDonnell Douglas, upgraded to the Dragon II in 1985, and then to the Super Dragon in 1990. The Dragon missile weighs approximately 14.5kg (32lb) and is 0.85m (33.3in) in length. Its range is nearly 1006m (1100yds). The weapon's major drawback was that the operator was required to remain in position and guide the missile to its target during flight. In recent years, the Dragon has been replaced by the FGM-148 Javelin.

During the Cold War, France entered into a joint

The French-designed ENTAC anti-tank missile was deployed in 1963 and could deliver a 4kg (8.8lb) warhead. French troops are spotting targets in the background.

development programme with Germany and Canada to produce the HOT I (High Subsonic Optical Guided) anti-tank missile system. Developed by Euromissile, now known as MBDA, the HOT I was introduced in 1978, weighs 24kg (52.9lb) and is 1.27m (4.1ft) long. Powered by a two-stage solid rocket motor, its maximum range is nearly 4000m (4375yds). Similar to the TOW, the HOT missile is an optical wire-guided system, which can be shoulder-fired or mounted on a vehicle or helicopter.

Eyeball to eyeball

During a briefing of Secretary of Defence Robert McNamara and the Joint Chiefs of Staff on

1 October 1962, a report revealed the potential presence of Soviet-made, intermediate-range ballistic missiles in Cuba:

'The source of this report stated that on 12 September he had personally seen some 20 such missiles in the vicinity of Campo Libertad, a small airfield on the western edge of Havana. While this report is still unconfirmed and there are no other reports concerning the presence of either SS-3 or SS-4 missiles, it is significant to note that by using the approximate center of the restricted area referred to above as a point of origin and with the radius of 1100nm [nautical miles], the accepted range of the SS-4 missile, the arc includes the cities of Philadelphia, Pittsburgh, St. Louis, Oklahoma City, Fort Worth-Dallas, Houston, San Antonio, Mexico City, all of the capitals of the Central American nations, the Panama Canal, and the oil fields in Maricaibo, Venezuela. The presence of operational SS-4 missiles in this location would give the Soviets a great military asset.'

The Cuban Missile Crisis brought the world to the brink of nuclear war. Although the presence of intermediate-range ballistic missiles constituted a strategic threat, another threat also existed in the form of Soviet tactical missiles armed with nuclear warheads, deployed in Cuba and serviced by Red Army advisors. President John F. Kennedy considered an invasion of Cuba by US troops. Had the order been given, the Americans would likely have been met by hostile nuclear fire.

The tactical nuclear rockets that American troops would probably have encountered belonged

Right: A single soldier fires the US M47 Dragon medium-range anti-tank missile. Lightweight and highly mobile, the Dragon delivers a warhead weighing 5.4kg (11.9lb).

HEAD TO HEAD: *US Dragon ATGW* VERSUS

Manufactured by McDonnell Douglas Corporation and introduced into the US military in 1970, the M47 Dragon anti-tank missile has undergone at least two revisions during its service life and has become known as the Superdragon in later versions. Its warhead is capable of penetrating armour up to 500mm (19.6in) thick.

US Dragon ATGW

Length: 74.4cm (29.3in)
Finspan: 34cm (13.4in)
Diameter: 25cm (9.8in)
Weight: 10.9kg (24lb)
Speed: 90m/sec (300ft/sec)
Range: 1000m (3300ft)
Propulsion: array of small solid-fuelled rocket motors;
    1.2kN (265lb) each
Warhead: 2.5kg (5.5lb) shaped charge

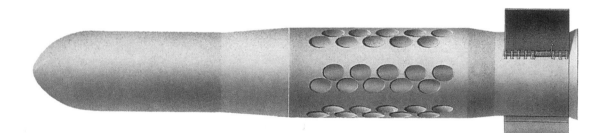

STRENGTHS

• Excellent mobility
• Expendable launcher
• Penetrating power

WEAKNESSES

• Firing signature
• Limited range
• General inaccuracy

Soviet Sagger AT-3

Deployed with the Soviet Red Army in 1961, the 9M14M Malutka, which translates as 'little baby', this wire-guided anti-tank missile was called the Sagger AT-3 by NATO personnel. The missile was fired by infantry or mounted atop a vehicle.

Soviet Sagger AT-3

Weight: 10.9kg (24lb)
Length: 860mm (33.86in)
Width (wingspan): 393mm (15.5in)
Diameter: 125mm (4.9in)
Effective range: 500–3000m (1640–9842ft)
Warhead: 2.5kg (5.5lb); HEAT 400mm (15.75in)
    versus RHA
Speed: average: 115m/sec (9377ft/sec);
    maximum: 200m/sec (656ft/sec)

STRENGTHS

- Excellent firepower
- High mobility
- Rapid deployment

WEAKNESSES

- Visual tracking
- Firing signature
- Slow speed

Shown atop its light tracked launcher, the Soviet Frog 5 missile is capable of launching within 40 minutes and reloading in little more than one hour.

to the series of mobile rockets known to NATO as FROG (Free Rocket Over Ground) and to the Soviets as Filin, Mars, Luna, Luna-1 and Luna-M. The first of these had entered service in the mid-1950s, armed with nuclear warheads. Those deployed in Cuba were likely FROG 1, 2 or 3 variants. Of these, the nuclear-armed FROG-1 was just over 10.4m (34ft) long with a fully armed weight of 4930kg (10,869lb) and a range of 25km (15.5 miles). Eventually, seven variants of the FROG were produced. The FROG 7 entered Red Army service in 1965 and is still in use among the countries of the former Soviet Union. Its length is nearly 9.5m (31ft), and its spin-stabilized maximum unguided range is 967km (601 miles).

In 1976, the Soviets began a gradual deployment of the Tochka (Point) tactical ballistic missile system, which combines the OTR-21 launch system and the 9M79 missile, as a replacement for the FROG series. Referred to by NATO as the SS-21 Scarab, the A and B variants of the Tochka system are capable of firing high-explosive, chemical, submunition (B only) and nuclear warheads. The length of both is 6.4m (21ft), and the launch weight is virtually identical at 2000kg (4409lb). The single warhead payload is 482kg (1063lb). The weapon is carried on a multi-wheeled transporter vehicle designated BAZ 5921 or ZIL-5937, including the erector and launcher. An accompanying transloader carries two additional missiles.

Shortly after World War II, the Soviet Union introduced the BM-14 140mm multiple rocket launch system, which continued the lineage of the

famous Katyusha rocket and was capable of firing 16 tube launched rockets affixed to a truck chassis. The rockets carried an 8kg (17.6lb) warhead with a range of just over 9.7km (6 miles). Nearly half a century following its introduction, the BM-14 and similar systems are still in use today, primarily in guerrilla-warfare situations. The BM-24 system was also a post-war truck mounted weapon and could fire 12 rockets with warheads up to 27.4kg (60.4lb) a distance of more than 10.4km (6.5 miles).

In the 1960s, the Red Army deployed the 9K51, also known as the BM-21 Grad (Hail), 122mm multiple launch rocket system. The BM-21 replaced the BM-14 system, and its numerous variants are capable of firing 40 rockets of just under 3m (10ft) in length with warheads up to 25kg (55.1lb) a maximum range of nearly 40.25km (25 miles). An entire salvo of rockets can be fired in the span of 20 seconds. The 220mm 9P140 Uragan (Hurricane), called the BM-27 by

NATO, was originally deployed in the late 1970s. Its 16 rockets may be fired as a salvo within 20 seconds and are carried in tubes mounted aboard a ZIL-135 eight-wheeled transporter vehicle. Powered by two petrol engines, the rockets deliver high-explosive, chemical or submunition warheads weighing up to 100kg (220.5lb) a maximum range of 35km (21.7 miles). The 9K58 Smerch (Tornado) is known to NATO as the BM-30. Its 12 rockets, with warheads weighing up to 258kg (568.8lb), may be fired within 38 seconds with a maximum range of almost 90km (56 miles).

The most widely known NATO multiple launch rocket system is the M270 MLRS, which was first deployed in 1983. A design collaboration between the United States, Great Britain, Germany and France, the system is affixed to the chassis of the Bradley Fighting Vehicle and is serviced by a crew of three. The M270 and the M270A1 variant are capable of firing a variety of rockets armed with high explosive, chemical, or submunition warheads. The system's 12 rockets may be fired in less a minute, while it is also capable of launching a pair of tactical missiles in 20 seconds. Guided and unguided projectiles have a range of 42km (26 miles), while the weaponry of the US Army Tactical Missile System (ATACMS) extends the range to 351km (186 miles). The M270 MLRS was originally known as the General Support Rocket System (GSRS) and renamed in 1979.

Following the deployment of the T34 series, along with the T39 and T40/M17, during World War II, interest in rocket artillery faded in the United States, and the M270 was the first system deployed by the US Army in approximately 35 years. Production of the M270 ended in 2003, and the system is in the service of several countries outside NATO.

Representative of the rocket artillery of the Cold War era People's Republic of China is the Type 63/81 107mm Multiple Launch Rocket, which entered service with the People's Liberation Army in 1963. Serviced by a crew of five, the system fires up to 12 rockets with an 8.3kg (18.3lb) warhead a distance of 8.5km (5.3 miles). A full barrage may be fired in less than nine seconds, and the reloading time is three minutes. Although it was being phased out of service with the Chinese military by the 1990s, it remains on duty with a number of countries.

Other Cold War rocket artillery systems include the French Lance-Roquettes Multiple (LRM), which is capable of firing 12 rockets or two missiles; India's Pinaka launcher, developed in the early 1980s and capable of firing 12 high-explosive rockets with warheads of 100kg (220.5lb) up to a distance of 40km (24.9 miles); and the Brazilian Astros II MLRS, which was deployed to the Brazilian Army in 1983 and is in service today with several Middle Eastern countries. The Astros II and its variants fire rockets of 127mm, 180mm and 300mm.

The Soviet SS-21 short-range tactical ballistic missile is seen atop its 9P129 wheeled transport and erector system. The SS-21 replaced the Frog system in the 1980s.

Left: The Astros II MLRS, a multiple launch rocket system, was developed in Brazil and entered service with that nation's military in 1983.

Ring of Fire

From 1965 to 1972, a total of 95 Soviet air defence missile systems had been delivered to North Vietnam, including 7658 missiles. By 27 January 1973, when the Paris Peace Accords ended US military involvement in the Vietnam War, more than 6800 missiles had been fired against US aircraft. Most of these had been the DVINA or S-75, known as the SA-2 Guideline to NATO. The Vietnam War was a proving ground for both surface-to-air missile technology and the countermeasures, from evasive tactics to sophisticated radar-jamming devices, which were available at the time. During the course of the war, Hanoi, the capital of North Vietnam, became the focus of heavy US air raids and was defended by hundreds of anti-aircraft guns and missiles.

In Air University Review, Colonel James H. Kasler, operations officer of the US 354th Tactical Fighter Squadron based in Thailand, recalled a memorable air strike on 29 June 1966 against a POL (petroleum, oil and lubricant) storage facility located just outside Hanoi:

'We spent six hours planning, checking, and double-checking every facet of the mission. This was our first detailed study of the defenses in the Hanoi area, and we found little in the aerial photographs to give us comfort. The enemy's air defenses, formidable from the start, were becoming more formidable each day. By every estimate, Hanoi had the greatest concentration of anti-aircraft weapons ever known in the history of aerial warfare. In Vietnam itself, there were from 7,000 to 10,000 fast-firing anti-aircraft weapons of 37mm caliber or larger. In addition, the Russians had provided the Vietnamese with a sophisticated radar and communications network for detection and coordination of their surface-to-air missiles (SAM) and MIG fighters.'

Although anti-aircraft guns employed by the North Vietnamese were often of World War II vintage, some had been updated as well. The missiles, however, were a different story. They were modern, radar-controlled and deadly. Pilots often referred to the big projectiles as 'flying telephone poles', and many aircraft were lost to the SA-2 during the war. Colonel Kasler flew the Republic F-105 Thunderchief fighter bomber:

'The 355th struck from the north. The plan was to cross the Red River 100 miles northwest of Hanoi, turn east, and descend to low altitude to avoid SAM missiles. Our route took us parallel and north of Thud Ridge, the 5,000-foot razorback mountain running west to east through the heart of North Vietnam. The eastern tip of the mountain ended about 25 miles due north of Hanoi. We would screen ourselves behind the mountain until we reached the eastern tip, then make a 90° turn south toward Hanoi.

'When we passed our initial point at the end of Thud Ridge, I called the flight to push it up and started a turn south toward Hanoi. As we turned, the fog bank faded away beneath us and we broke

Seen in tow behind a truck, the Soviet-made SA-2 was a primary air defence missile supplied to North Vietnam. Nearly 7000 were fired at US aircraft in Vietnam.

HEAD TO HEAD: *BM-21* VERSUS

The BM-21 Grad, or 'Hail', was the mobile truck platform for the Soviet 122mm multiple launch rocket system developed during the early 1960s. More than four decades later, the Grad remains a frontline system with the armed forces of over 40 nations.

BM-21

Crew: 6
Weight: 13,700kg (30,203lb)
Dimensions: length: 7.5m (24.1ft); width: 2.69m
 (8.83ft); height: 2.85m (9.35ft)
Range: 1000km (621 miles)
Armour: n/a
Armament: 122mm (4.8in) type in different lengths
 with option of several warheads including 45 anti-
 tank submunitions
Powerplant: one water-cooled 134kW (180hp)
 V-8 petrol engine
Performance: maximum road speed: 80km/h
 (50mph); fording: 1.5m (4.9ft); vertical obstacle:
 0.65m (2.13ft); trench: 0.875m (2.87ft)

STRENGTHS	WEAKNESSES
• Rapid deployment	• Somewhat inaccurate
• Concentrated firepower	• Little crew protection
• High mobility	• Noticeable signature

US MLRS

The M270 MLRS multiple launch rocket system was designed jointly by the United States, Great Britain, Germany and France during the 1970s and placed in service in 1983. The missiles are contained in interchangeable pods, while the engine is a 298kW (400hp) diesel.

US MLRS

Crew: 3
Weight: 24,721kg (54,500lbs)
Dimensions: length: 6.83m (22.4ft); width: 2.99m
    (9.8ft); height: 2.59m (8.5ft)
Operational range: 483km (300 miles)
Rate of fire: rockets: 12in less than 60sec;
    missiles: 2 in 20sec
Primary armament: M269 Launcher Loader Module
Engine: Cummins Diesel 298kW (400hp)
Speed: 64km/h (40mph)

STRENGTHS

• Shoot and scoot
• Varied ordnance
• Rapid deployment

WEAKNESSES

• Complex loading
• Noticeable signature
• Tracked chassis

With a Lockheed U-2 spy plane in the background, pilot Francis Gary Powers (right), who was shot down over the Soviet Union in 1960, discusses flight operations.

into the clear. At that same instant, flak began bursting around us. I glanced to the right toward Phuc Yen airfield and could see the flak guns blinking at us. Despite the fact that we were only 300 feet above the ground, the Vietnamese had leveled their heavy 85mm and 100mm guns and were firing almost horizontally at us… As I was turning in, I could see three 10-gun 85mm batteries on Gia Lam airfield frantically firing. Ignoring these as best I could, I began my bomb run. Amazingly, only one strike aircraft was lost to flak in the raid; the pilot, Captain Neil Murphy Jones, was interned in North Vietnam until February 1973.

'One of the puzzles of the raid was why the Vietnamese had not fired any of the dozens of SAM missiles that rimmed Hanoi. The day following the raid, they began firing SAMs in volleys at our aircraft, which was a complete change in the tactics they had used previously. The answer was learned two months later when I was shot down and captured by the North Vietnamese.

'Shortly after my capture on 8 August 1966, I was questioned by a Vietnamese interrogator while lying in a hospital room in Hanoi. The interrogator tried to get information from me concerning the Hanoi POL strike. He asked: "What did you think about our defenses during the Hanoi raid?" I said, "I figure you got a new air defense boss." Just a guess on my part, but apparently a correct one as he became quite agitated and left. A short time later my room was invaded by four very stern-looking Vietnamese, who spent the next two days trying to

figure out how I knew they had a shake-up in their air defense command.'

Anti-aircraft guns

Nazi Germany developed the first anti-aircraft missile, the Wasserfall, or Waterfall, during World War II. The remote-controlled rocket engine of the Wasserfall was powered by liquid fuel and successfully lifted the weapon skywards in February

1945, just three months prior to the end of the war. When hostilities ceased, the United States and the Soviet Union scrambled to collect German military hardware and technology, and high on the list of priorities was the Wasserfall. Both nations used the German technology as the basis for early anti-aircraft defence missile systems of their own.

The Soviet S-25 Berkut (Golden Eagle), which was known to NATO as the SA-1 Guild, became the first operational surface-to-air missile. Development commenced on 9 August 1950, and test-firing was completed in January 1952. By March 1954, the SA-1 was being delivered to the Red Army for further testing, and in June 1956 the weapon became combat operational. The SA-1 was the progenitor of numerous variants, which remained in service until their retirement in 1982 in favour of the modern S-300 system. In 1956, the Soviet capital of Moscow was protected by two rings of 56 SA-1 launch sites, totalling more than 3000 missiles. The V-300 variant of the SA-1 was fired initially on 25 July 1951. Carrying a warhead of up to 318kg (700lb), the missile could reach altitudes of 914–18,288m (3000–60,000ft).

Perhaps the most famous of the Soviet anti-aircraft missiles is the S-75, developed by Lavochkin, which reached field units of the Red Army in 1957 and was later exported to a number of countries. Also known as the DVINA, the missile was designated the SA-2 Guideline by NATO. The SA-2 was originally conceived as a replacement for a number of anti-aircraft guns, including batteries of 130mm and 100mm weapons. The two-stage missile delivers a fragmentation warhead of up to 195kg (430lb) to a maximum altitude of 18,288m (60,000ft). Its operational range is 48km (30 miles), and the warhead's lethal radius is up to 250m (820ft).

It was an SA-2 that shot down American U-2 pilot Francis Gary Powers in 1960 while he was on a surveillance mission high above a testing site near Sverdlovsk. Aside from the extensive use of the SA-2 against US aircraft during the Vietnam War, variants of the missile system remain in service with at least 35 nations today, including Vietnam, Egypt, Poland and North Korea.

While the SA-2 system initially required a fixed launch site, the Isayev S-125 Neva/Pechora, known to NATO as the SA-3 Goa, complemented the SA-2 as a somewhat mobile short-range missile, which could be launched from fixed sites or from the bed of a ZIL truck. Originally placed in service between 1961 and 1964, the SA-3 was withheld from service by the Soviets during the Vietnam War due to the fact that they feared the Chinese would attempt to copy its design. The original SA-3 was capable of reaching altitudes greater than 10,000m (32,800ft) with a range of up to 15km (9.3 miles). Although it is considered obsolete today, the system has been upgraded numerous times. A modified SA-3 actually shot down a US F-117 Nighthawk stealth fighter in Kosovo on 27 March 1999.

The 9M311, called the SA-19 Grisom by NATO, is a two-stage air-defence missile launched from the mobile Kashtan/Kortik and Tunguska systems. The SACLOS-guided Grisom weighs 57kg (125.7lb) and delivers a 9kg (19.8lb) warhead to an altitude of 3500m (11,483ft). The warhead is triggered by a laser fuse, which detonates when the projectile is within 5m (16.4ft) of its target.

The S-300 system, called the SA-10 Grumble by NATO, was introduced in the late 1970s, and within a decade more than 80 launch sites were scattered throughout the Soviet Union. Upgrades have taken place periodically during the last 25 years, and the S-300 system has become one of the most lethal in the world. The original version, the S-300PT, combined a large warhead, high manoeuvrability, long range and extremely sensitive low-altitude radar, placing the weapon in a class all its own: it has the ability to engage multiple targets

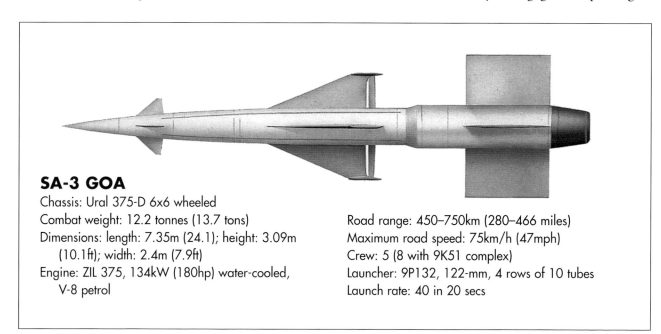

SA-3 GOA

Chassis: Ural 375-D 6x6 wheeled
Combat weight: 12.2 tonnes (13.7 tons)
Dimensions: length: 7.35m (24.1); height: 3.09m (10.1ft); width: 2.4m (7.9ft)
Engine: ZIL 375, 134kW (180hp) water-cooled, V-8 petrol

Road range: 450–750km (280–466 miles)
Maximum road speed: 75km/h (47mph)
Crew: 5 (8 with 9K51 complex)
Launcher: 9P132, 122-mm, 4 rows of 10 tubes
Launch rate: 40 in 20 secs

in the same fire-control system. Depending on type, the warheads may weigh up to 150kg (330lb), and the missiles themselves as much as 1801kg (3970lb). The range of the various missiles is up to 402km (250 miles), and one variant has been equipped with a nuclear warhead.

Another Soviet surface-to-air missile worthy of note is the 9K38 Igla (Needle), known to NATO as the SA-18 Grouse. A shoulder-fired missile with a two-colour infrared guidance system capable of delivering a 1.2kg (2.65lb) warhead to an altitude of 3500m (3827yds) at a range of 5200m (5687yds), the Igla was deployed with the Red Army in 1983 after 12 years of development. During the process, it had been recognized that the development of a fully capable Igla would require an inordinate amount of time, and the SA-16 Gimlet, with limited capability, was issued to Soviet troops in 1981.

Bristol Bloodhound

The primary British surface-to-air missile of the Cold War era was the Bristol Bloodhound, which was developed during the 1950s and entered service with the Royal Air Force in 1958. The Bloodhound served until 1991, when the Mark II variant, introduced in 1964, was retired. The Bloodhound line began as the Red Duster Project during the late 1940s, and nearly 800 of the missiles, weighing in excess of 2265kg (5000lb) with a range of nearly 185km (115 miles), were manufactured. A nuclear-capable Mark III and mobile Mark IV programme were both cancelled.

The Bristol Bloodhound was designed during the 1950s and served for more than 30 years. The pictured Mark II Bloodhound countered high-altitude bombers.

An early example of British air-defence missile technology, the English Electric Thunderbird was conceived in the late 1940s as a mobile system for use with troops in the field.

The English Electric Thunderbird, capable of delivering a high-explosive warhead a maximum distance of more than 74km (46 miles), was introduced to the British Army in 1959. The Thunderbird was originally guided by AMES Type 83 pulsed radar. Later, the weapon was improved with AMES Type 86 continuous wave radar and designated the Thunderbird II. The Thunderbird was the first British designed and manufactured missile to serve with the nation's armed forces. The original Thunderbird was withdrawn in 1963, and the Thunderbird II was retired in 1976.

Developed in the 1960s, the solid rocket propelled Rapier missile is capable of delivering a fragmentation warhead to an altitude of 3000m (9842.5ft) and a range of 6.8km (4.2 miles). The Rapier entered service with the Royal Air Force in

1971, and upgrades have included a mobile tracked version, which requires approximately 30 seconds for deployment into firing mode. The Rapier was used in action during the Falklands War of 1982 and claimed 14 confirmed kills and six probables. Development of the current version, the Rapier FSC, was begun in the 1980s, and the system is currently in use with the Royal Artillery.

During the 1950s, the British conducted the Green Cheese missile project, testing a naval missile capable of carrying a nuclear warhead. Green Cheese was scrapped in 1956 due to its extreme weight of 1724kg (3800lb) and large cost overruns. The British also considered development of three anti-aircraft guns during the Cold War, including the rapid-firing Green Mace 5in gun, a 4in variant of the Green Mace capable of firing up to 96 rounds per minute, and the 42mm Red Queen developed by the Swedish firm of Oerlikon to replace the antiquated Bofors 40mm gun.

Each of these programmes was cancelled in favour of missile projects.

Anti-aircraft missile project

In the spring of 1945, the United States embarked on the development of its first anti-aircraft missile project, Nike. The result was the nation's first operational anti-aircraft missile system, the Nike Ajax. The Ajax was successfully tested in November 1951, and went into active service two years later. By 1962, more than 300 launch sites had been positioned to protect high-value potential targets in the United States and replaced nearly 900 radar equipped anti-aircraft guns. The Ajax was manufactured by Douglas and capable of reaching an altitude of 21,336m (70,000ft) and a range of 40km (25 miles). Its three explosive charges were designed to increase the probability of a lethal strike.

The next generation Nike missile, the Hercules, increased the range to 160km (100 miles) and altitude to about 30,480m (100,000ft). It was normally armed with a fragmentation warhead but was also nuclear capable. According to records, the Hercules was deployed around Chicago, Illinois, in the summer of 1958, and within two years 88 Hercules batteries had joined 174 Ajax batteries in 23 defensive zones in 30 states. The deployment of the Hercules peaked in 1963 at 134 batteries. The advent of intercontinental ballistic missiles and SALT treaty agreements with the Soviet Union rendered the Ajax and Hercules systems obsolete by 1974. An improved Hercules and the Nike Zeus missiles were researched between 1959 and 1963 but cancelled completely by 1966. Today, a few Nike launch sites have been preserved as museums.

In 1960, the Raytheon MIM-23 HAWK (Homing All the Way Killer) was introduced in the US armed forces as a medium-range surface-to-air missile. It was eclipsed by the MIM-104 Patriot system in the mid-1990s, but it was not retired from service until 2002. The HAWK was mobile and similar in performance to the Soviet SA-3 and SA-6 systems. Tracing its development to 1952, the Hawk, which was serviced by a battery crew of 51 soldiers, delivered a 54kg (119lb) fragmentation warhead to an altitude of 13,716m (45,000ft) and a range of 24km (15 miles). The missile has seen extensive combat use. During the Yom Kippur War, Israeli HAWKs downed as many as 24 Egyptian aircraft, and Iranian HAWK missiles were reported to have downed 40 Iraqi planes during the Iran-Iraq War of the 1980s. HAWK missiles deployed by the Kuwaiti armed forces shot down seven Iraqi aircraft and one helicopter in 1990.

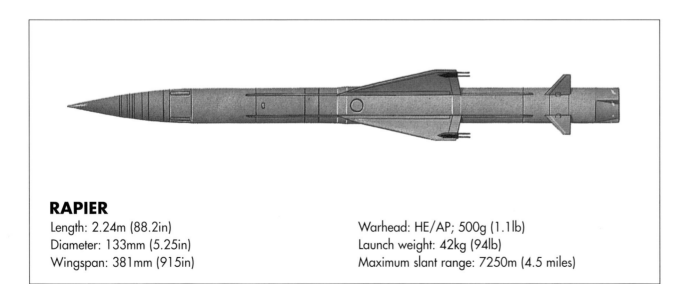

RAPIER
Length: 2.24m (88.2in)
Diameter: 133mm (5.25in)
Wingspan: 381mm (915in)

Warhead: HE/AP; 500g (1.1lb)
Launch weight: 42kg (94lb)
Maximum slant range: 7250m (4.5 miles)

The self-propelled MIM-72/M48 Chaparral missile system became operational with the US Army in 1969 and was fully retired from service by 1998. The Chaparral utilized the MIM-72A missile, a variant of the highly successful AIM-9D Sidewinder air-to-air missile, with slight modifications to its stabilizing fins. Employing an infrared guidance system, the missile delivered an 11kg (24.3lb) warhead to a maximum altitude of 3000m (9842ft).

The MIM-104 Patriot surface-to-air missile system embodies air-defence capabilities primarily in the ABM role, intercepting enemy missiles in the air. It has also been the replacement system for both the Nike Hercules and HAWK systems. Research on the Patriot system began in the 1960s, and by 1975 a successful interception test had taken place. The following year, the programme, originally known as the SAM-D, was renamed the Patriot Air Defense Missile System. In 1988, the scope of the Patriot system was expanded to include not only defence against aircraft but also against missiles. Fired from highly mobile units mounted on M860 trailers that are towed by the M983 HEMTT 'Dragon Wagon', the Patriot can be readied from mobile to firing configuration in about 45 minutes. The system gained fame during Operation Desert Storm in defence of Israel and Coalition airspace against Iraqi SCUD missiles.

The Patriot fires versions of eight different missiles. Of these, the PAC-2 is propelled by a solid fueled rocket engine, weighs 900kg (1984lb), and has a range of 160km (99 miles). Upgrades involving computer programs and software, advanced radar and missile design have taken place during the 1980s and 1990s.

High Chaparral

The American-built MIM-72/M48 Chaparral air-defence system entered service in the late 1960s. The MIM-72 missile was actually adapted as a mobile surface-to-air version of the US Navy's highly successful AIM-9 Sidewinder air-to-air missile. It is shown mounted atop a variant of the M113 chassis. The MIM-72 was intended as a long-range complement to the short-range Vulcan system, and was retired from service during the 1990s.

The shoulder-fired FIM-92 Stinger missile was placed in service by the United States in 1981, generally replacing the Redeye missile, which was tested in 1960 and operational with US forces from 1969 to 1985. The Stinger utilizes a passive infrared homing guidance system and delivers a 3kg (6.6lb) warhead at a maximum altitude of 4800m (15,750ft). Manufactured by Raytheon, the weapon is also produced under licence by the German firm EADS.

Since the inception of the programme, more than 70,000 Stinger missiles have been deployed with the armed forces of up to 30 nations. The weapon is easy to transport and may be fired by a single soldier. The entire missile-launcher package weighs only 15kg (33.5lb). The Stinger was used on a limited basis by British forces during the Falklands War in 1982 and shot down an Argentine aircraft. However, the most publicized use of the Stinger to date has been with the Mujahideen guerrillas opposing Soviet forces in Afghanistan during the 1980s. Reportedly, the Afghan warriors scored a number of successes against Soviet planes and helicopters.

France and Germany cooperated on the development of the Roland surface-to-air missile during the 1970s, and the weapon was modified with improved radar, extended range and a larger warhead during the 1980s. Sill in service with numerous countries, the Roland generally delivers a 6.5kg (14.3lb) pre-fragmentation high explosive warhead to an altitude of more than 5486m (18,000ft) at a range of 7.9km (4.97 miles).

The US MIM-104 Patriot surface-to-air missile system gained fame during Operation Desert Storm firing anti-ballistic missile missiles against Iraqi Scuds.

The Chinese Kai Shan 1 missile system was developed in the 1980s along similar specifications to those of the US HAWK. Tested successfully in 1989 and debuted at the Zhuhai Air Show in 1998, the weapon carries a 100kg (220.5lb) warhead a maximum altitude of 24,000m (78,740ft) at a range of 7km (4.35 miles).

The end of superpower confrontation

Although the image of the Berlin Wall crumbling under the blows of sledgehammers is indelibly etched in the minds of those who recall the event, the end of the Cold War did not signal the end of strife, discord and armed conflict around the world. The current war against global terrorism is unlike any conflict ever prosecuted. It has no defined front line. It has no uniformed enemies fighting linear battles. It knows no distinction between civilian and military.

Modern warfare is certainly conducted with modern arms, advanced technology and smart weapons. Still, those who are aware that they cannot compete on the basis of cutting-edge technology have resorted to low technology such as the improvised explosive device (IED), the car bomb and the sniper rifle.

In today's warfare, Technology rules the battlefield, but only to the extent that it can be successfully employed. When the enemy is concentrated and his position may be found and fixed, the firepower of modern artillery swings into action with devastating results.

Shown hurtling from its wheeled launcher, the Roland surface-to-air missile was the product of a collaboration between French and German designers.

MODERN ARTILLERY

Perhaps no other period in history has witnessed more rapid advances in battlefield technology than the last 150 years. With an ever-quickening pace, developments in multiple weapons systems are continually pushing the frontiers of military capability. Artillery remains an integral component of modern combat, delivering the hammer blow in support of infantry.

As authors Richard A. Gabriel and Karen S. Metz explain: 'Modern artillery is lighter, stronger, and more mobile than ever before. Computerized fire direction centres can range guns on target in only 15 seconds compared to six minutes required in World War II. The rates of fire of these guns are three times what they used to be. So durable are the new artillery guns that they can fire 500 rounds over a four-hour period without incurring damage to the barrel. Range has increased to the point where the M-110 gun can fire a 203mm shell 25 miles. The self-propelled gun has a travel range of 220 miles at a speed of 35 miles per hour. Area saturation artillery, in its infancy in World War II, has become very lethal. A single Soviet artillery battalion firing 18 BM-21 rocket launchers can place 35 tons of explosive rockets on a target 17 miles away in just 30 seconds. The

Left: An American M198 155mm howitzer is fired at Iraqi positions while its crewmen cover their ears to protect against the weapon's distinctive report.

American Multiple Launch Rocket System (MLRS) is a totally mobile self-contained artillery system that can place 8000 M-77 explosive rounds on a target the size of six football fields in less than 45 seconds. Air defence guns have developed to where a single M-163 Vulcan cannon can fire 3000 rounds of explosive 20mm shot per minute with almost 100 percent accuracy within two miles of the gun position. Modern anti-aircraft guns command 36 times the airspace around their position as they did in World War II.'

The modern battlefield remains a complex arena not only due to advances in technology, but also to the nature of combat itself during the post-Cold War era. While Coalition forces have opposed the Iraqi military during Operation Desert Shield/Storm and Operation Iraqi Freedom, numerous conflicts with organized guerrilla movements have erupted around the world. The United States, Great Britain, Pakistan and NATO forces have battled the Islamic fundamentalist Taliban and terrorist Al Qaeda in Iraq, Afghanistan

and the rugged mountains of the Afghan-Pakistani border. Russia has also fought two difficult wars in the breakaway region of Chechnya. Centuries of ethnic animosity in the Balkans have boiled over into open conflict in Bosnia and Kosovo. The horn of Africa continues to suffer in a cycle of violence, a flame that was fuelled by the rivalry between the superpowers during the Cold War.

During each of these conflicts, whether the warring parties are organized military formations or bands of irregulars fighting for control of certain territory and its indigenous population or to perpetuate a pseudo-theological agenda, the role of artillery has been prominent. Guns and howitzers remain capable of inflicting heavy casualties on enemy troops, as well as striking terror into a civilian population and destroying the infrastructure of a major city. Rockets, anti-tank missiles and anti-aircraft systems have become increasingly sophisticated and deadly. The development and availability of tactical ballistic missiles have raised the stakes considerably as well.

In the hands of those intent on utilizing them, tactical ballistic missiles may deliver conventional, chemical or nuclear warheads. The collapse of the Soviet Union and the rise of a significant international arms trade have made such a prospect frighteningly real.

Scud versus Patriot

During Operation Desert Storm, Iraqi dictator Saddam Hussein ordered multiple attacks against targets in Kuwait, Saudi Arabia and Israel with short-range ballistic missiles of the Soviet-made Scud family. Although the prospect of a nuclear attack was thoroughly discounted, it was known that the Iraqis possessed chemical weapons. To counter the Scud threat, the United States deployed the Patriot anti-ballistic missile (ABM) system. The Scud attacks against Israel nearly compelled the Jewish state to enter the Gulf War. Gas masks were distributed to the Israeli population, Israeli Air

The M110 8in (203mm) self-propelled howitzer is the largest weapon of its type in the US arsenal. The system entered service in 1963 and has officially been retired.

Force pilots remained on full alert, sometimes sleeping strapped into their cockpits, and other measures were taken to safeguard against a chemical attack.

Major General Ya'acov Lapidot, director-general of the Israeli Ministry of Police, wrote:

the treaty terms and provided Spiders to Eastern European countries.

With a range of 483km (300 miles) and a somewhat larger warhead than the R-17 at 699kg (1540lb), the 9K714B Tochka tactical ballistic missile appeared with Red Army units in 1976. Essentially the same missile configuration as the SS-23 Spider, the Tochka first flew in 1974 and was distinguished by its ZIL-375 transporter rather than the eight-wheeled TEL which carried the SS-23.

In 1999, the SS-26 Iskander entered service with the Russian Army and constituted a second replacement for the elderly Scud. Development of the SS-26 was begun in the 1980s, and the first successful firing occurred in October 1995. Highly mobile, the Iskander is deployed in pairs. Each missile contains a warhead of 480kg (1,058.2lb), weighs 3800kg (8,377.6lb), and is capable of hitting targets 280km (174 miles) distant. The Iskander is controlled by a satellite terminal guidance system and correcting and target locking technology.

The first fully operational, nuclear-capable, surface-to-surface rocket in the US arsenal was the MGR-1 Honest John, of which more than 7000 were produced from 1951 to 1965. The Honest John was initially deployed with US forces in 1953 and designated the M31. The M31 was mobile in three sections, which could be assembled and readied to fire in as little as five minutes, including time for mounting on an M289 launcher. Initially, the rocket was fitted with a W7 nuclear warhead capable of delivering a yield up to 20 kilotons. Later, this was increased with the W31 warhead and its yield of up to 40 kilotons. Warheads containing lethal Sarin nerve gas were also configured for the Honest John by 1960, and

Soldiers sift through the rubble of a building demolished by an Iraqi Scud missile. Scuds caused the greatest loss of life among American military personnel during the Gulf War.

the system's maximum range was 25km (15 miles). In 1973, the MGR-1 was supplanted by the MGM-2 Lance missile system with the US Army; however, it remained in service with NATO forces until 1982.

The MGM-5 Corporal missile system, with a range of 139km (86 miles), was deployed to US troops in Europe in 1955. Capable of firing conventional high explosive or nuclear warheads, the Corporal was developed at the White Sands Missile Range and built upon the technology of the German V-2 rocket of World War II. The weapon was exported to Great Britain in the 1950s and remained in service until 1964. The Corporal was relieved of duty by the MGM-29 Sergeant, a solid fuel missile developed by the US Jet Propulsion Laboratory. Armed with the W52 nuclear fission warhead with a payload of 200 kilotons, the

HEAD TO HEAD: *G-6* VERSUS

A 155mm howitzer of South African design and manufacture, the G-6 entered production with the LIW Division of DENEL Corporation in 1987. The weapon was built to fire ammunition that is also compatible with the earlier G-5 howitzer.

G-6

Crew: 6
Weight: 47,000kg (46.25lb)
Dimensions: length: 9.2m (30.2ft); width: 3.4m
    (11.2ft); height: 3.3m (10.8ft)
Range: 700km (435miles)
Armour: not disclosed
Armament: one 155mm (6.1in) gun
Powerplant: one diesel 392kW (525hp)
Performance: maximum road speed: 90km/h
    (56mph); fording: 1m (3.3ft); vertical obstacle:
    0.5m (1.6ft); trench: 1m (3.3ft)

STRENGTHS	WEAKNESSES
• High mobility	• High silhouette
• Excellent firepower	• Limited availability
• Mine protection	• Average range

Norinco PLZ45 155mm

Manufactured by China North Industries Group Corporation, the NORINCO 155mm gun howitzer is a self-propelled weapon developed for the burgeoning export market. In 1997, the PLZ 45 won the contract to supply weapons to the Kuwaiti Army over US and European competitors.

155/45 Norinco SP Gun

Crew: 5
Weight: 32,000kg (31.5lb)
Dimensions: length: 6.1m (20ft); width: 3.2m (10.5ft);
    height 2.59m (20ft)
Range: 450km (20 miles)
Armour: not disclosed
Armament: one 155mm (6.1in) gun
Powerplant: one diesel developing 391.4kW (525hp)
Performance: maximum speed: 56km/h (35mph);
    fording: 1.2m (4ft); vertical obstacle: 0.7m (28in);
    trench: 2.7m (8.9ft)

STRENGTHS

- High-performance chassis
- Rapid loading
- Competitive cost

WEAKNESSES

- Developed for export
- Western calibre ammunition
- Potential maintenance issues

Sergeant weighed 4581kg (10,100lb) and was 10.5m (34.4ft) long. Its maximum range was 139km (86 miles). Initially deployed with US forces in September 1962, the Sergeant was eventually replaced by the Lance system and completely retired from active service in the spring of 1977.

For 19 years, from 1973 to 1992, the MGM-52 Lance tactical ballistic missile was on station with US armed forces. Developed during the 1960s, the Lance was approved by the Department of Defense in the autumn of 1970. The Lance was armed with the W70 nuclear warhead, which could produce a payload of one to 100 kilotons; the W70-3 nuclear 'enhanced radiation' warhead, which was related to the neutron bomb; or the M251 conventional warhead with submunitions. Powered by liquid fuel, its range was 121km (75 miles), and its fully loaded weight was 1293kg (2850lb). The final Lance battalion was taken out of service in June 1992, and surplus rockets were used as targets for anti-missile systems testing. The FOTL (Follow On To Lance) missile project was planned with a range of 435km (270 miles), which is well beyond that of the older MGM-52, but it was terminated in the 1980s.

The current tactical ballistic missile system deployed by the United States is the Lockheed Martin MGM-140 ATACMS (Army Tactical Missile System), which was test-fired successfully in April, 1988, following six years of development. The MGM-140 may be launched from such mobile platforms as the MLRS M270 and M270A1 and the state-of-the-art HIMARS. Its Block II variant delivers the Grumman BAT (Brilliant Anti-Tank) guided submunition, while the Block I is generally armed with a warhead containing 950 M74 anti-personnel/anti-materiel

German scientist Wehrner von Braun was a pioneer in the development of rockets and ballistic missiles. A former SS officer, he was also implicated as a war criminal.

bomblets, which are scattered to an area of 33,445 square metres (360,000 square feet). Propelled by a solid fuel rocket motor, the ATACMS is approximately 30.5m (13ft) long and has a range of about 160km (100 miles). Each launcher holds two ready-to-fire missiles. During Operation Desert Storm, 32 ATACMS missiles were fired from the

M270 launcher, and during Operation Iraqi Freedom more than 450 have been launched.

A pair of tactical ballistic missiles have been produced by France during the last 35 years. The Pluton was developed as a replacement for the US Honest John and entered service in 1974. Its short range of 120km (74.6 miles) caused it to be replaced by the longer-range Hades in 1991. The Pluton was capable of delivering conventional high-explosive warheads or nuclear warheads of 15 and 25 kiloton payloads. The Pluton was fired from the TEL platform atop the AMX-30 tank chassis. The service life of the Hades was only five years, as the project was terminated in 1996 when the French government determined to pursue a defensive posture that did not emphasize land-based tactical ballistic missiles. The Hades carried either high explosive conventional or the TN-90 nuclear warhead with a payload of 80 kilotons. Its range was 480km (298.3 miles).

Dust in the wind

On 2 August 1990, the Iraqi leader Saddam Hussein launched the invasion of neighbouring Kuwait. In response, a multinational Coalition undertook a military buildup of several months, initiated an aerial bombardment campaign and ultimately ejected the Iraqis from Kuwait with a lightning fast 100-hour ground campaign. Operation Desert Shield/Storm bore witness to the awesome firepower and cutting edge military technology possessed by the Coalition forces. Iraqi military assets were routinely outgunned and outranged. The rapid deployment and movement of towed, anti-aircraft, anti-tank, rocket and self-propelled artillery proved decisive in the defeat of the Iraqis.

The only Iraqi ground offensive action of the Gulf War took place in January and February 2001, at the Saudi Arabian border town of Khafji. After the Iraqis occupied the town, Coalition forces systematically destroyed them, in part because of the information received from two artillery spotter teams, which had been inadvertently caught up in the Iraqi movement and remained there while Khafji was occupied.

Recalling his experience at Khafji, US Marine Captain Jim Braden remarked:

'We [used] Saudi artillery on the third day. The first two days, Marine artillery supported us. However, the Saudis insisted it was their sector and their responsibility, so they wanted to fire their artillery. They had the M109A2 155mm self-propelled howitzer. They would hit the target all right, but when we asked for them to cease fire, they would send in another barrage... I must say the forward observer ... was invaluable. He did a great job.'

A decade later, the United States and its allies are embroiled in a War on Terror following the Al Qaeda attacks of 11 September 2001. Once again, a multinational military Coalition, including troops under the auspices of NATO, has been engaged in Iraq and Afghanistan.

Commenting on the effectiveness of field artillery during Operation Iraqi Freedom, William G. Pitts wrote:

'The FA [field artillery] supporting manoeuvre forces during Operation Iraqi Freedom (OIF) proved to be the deciding factor in many of

the conflicts – although the enemy artillery outnumbered and outranged the Coalition Force FA. The FA in OIF was the lowest ratio of artillery pieces to troops in war since before World War I. Artillery fires came at a premium with lines of communication stretched from the Kuwait border to Baghdad, including ammunition supply...

'The magnificent soldier and Marine field artilleryman adapted to changes while rapidly moving great distances, made critical decisions independently in decentralized operations with little or no sleep and executed fire missions with extraordinary precision in constant movements-to-contact, meeting engagements and urban operations as part of the most effective joint fires

The successor to the infamous Scud-B missile in the Soviet arsenal, the SS-23, shown in transit, is a tactical theatre ballistic missile deployed near the end of the Cold War.

M109A6 PALADIN

Crew: 4
Weight: 28,738kg (29.28 tons)
Dimensions: length: 6.19m (20.3ftn); width: 3.15m
(10.3ft); height: 3.24m (10.6ft)
Range: 405km (252 miles)
Armour: not disclosed

Armament: one 155mm (6.1in) howitzer
Powerplant: one Detroit Diesel turbocharged two-
stroke 302kW (405hp) engine
Performance: maximum road speed: 56km/h
(35mph); fording: 1.95m (6.4ft); vertical obstacle:
0.53 (21in); trench: 1.83m (6ft)

team in history… The Army and Marine Field Artillery were key to combined arms operations and a major contributor to the joint fires team.'

According to Pitts, the Coalition M109A6 Paladin 155mm self-propelled howitzers were effective in a variety of missions along with the M198 155mm towed howitzer. The guns of the US Third Infantry Division alone fired nearly 14,000 rounds of 155mm ammunition, including 120 precision guided 'sense and destroy' armour munitions (SADARM). During heavy fighting at the town of Naseriyah, the 1/10 Marines of Task Force Tarawa fired more than 2000 155mm rounds. During Operation Iraqi Freedom, British forces fired over 9000 rounds of 155mm ammunition and over 13,000 105mm rounds.

Reporting on the performance of field artillery during the effort to clear insurgents from the Iraqi city of Fallujah, the *Washington Post* quoted one Marine as saying, 'It's made everybody get out of town'. The *Post* continued, 'Alpha Battery's two artillery pieces have fired more than 300 rounds in the first three days of the battle. The Marines' Mike Battery 414, which has six big guns at the same military outpost, has launched more than 500 rounds.'

The backbone of the self-propelled artillery in the US military for the last 40 years has been the M109 155mm howitzer family. The system was initially placed in service in 1963 and has been continually upgraded through the years. Currently, the M109A6 Paladin, manufactured by BAE Systems, is the frontline variant in use exclusively

by the United States. Mounting the M126 155mm howitzer, the Paladin can achieve a maximum rate of fire of four rounds per minute. Serviced by a crew of eight and capable of reaching a maximum speed of 56km/h (35mph), the system's secondary armament consists of a .50-calibre machine gun, Mk 19 Mod 3 40mm automatic grenade launcher or a 7.62mm M60 or M240 machine gun. Fully loaded, the M109 weighs 25 tonnes (27.5 tons).

The first M109s were produced in 1963, and the weapon initially saw combat during the Vietnam War. It was in action with Israeli forces during the Yom Kippur War of 1973, the conflicts in Lebanon, the Iran-Iraq War, the Gulf War of 1991 and Operation Iraqi Freedom. During Desert Storm, the M109 was deployed with the Saudi Arabian, Egyptian and British armies. Upgrades through the years have included improvements in reliability, availability and maintainability (RAM). The M109A6 Paladin exhibits stronger armour and an advanced inertial navigation system. Coordinated fire is achieved through the transmission of data to a battery fire direction centre. Approximately 1000 Paladins are currently in service with the US armed forces, while variants are in the arsenals of more than 20 other nations.

The XM2001 Crusader, intended as a successor to the M109, was originally scheduled for deployment in 2008. However, the $11 billion project was cancelled in 2002 after the secretary of defense, Donald Rumsfeld, concluded that the 39-tonne (43-ton) vehicle powered by an LV 100-5 turbine engine generating 1119 kW (1500hp)

Right: The projectile just fired from this M198 artillery piece near Fallujah, Iraq, can be seen at the top of this photograph. The crewmen are US Marines.

lacked the desired mobility and accuracy of fire. Therefore, the Paladin is expected to remain in service until a future generation of self-propelled weapons is developed within the US Army's Future Combat Systems (FCS) initiative.

During the mid-1990s, the British Army began replacing its complement of M109s with the 41-tonne (45-ton) AS-90 self-propelled gun. Developed in the mid-1980s, the AS-90 (Artillery System for the 1990s) is serviced by a crew of five. Mounting a 155mm cannon, the weapon is capable of a top speed of 55km/h (34mph). The main armament can fire at rates of three rounds in a 10-second burst, six rounds per minute for up to three minutes, or two rounds per minute for 60 minutes. Secondary armament consists of a 7.62mm L7 GPMG machine gun. Its powerplant is the Cummins VTA903T 492kW (660bhp) V8 turbodiesel engine. In 2002, a number of AS-90s were upgraded to a longer 52-calibre length cannon.

Representative of modern Russian self-propelled artillery are the 152mm 2S19 MSTA-S howitzer and the 120mm 2S31 Vena (Vein) gun. Development of the 2S19 began under the Soviet regime, and the weapon was deployed with the Russian Army in 1989. The high silhouette turret sits atop a chassis incorporating elements of the T-72 and T-80 heavy tanks. Weighing 37.5 tonnes (42 tons), the 2S19 is serviced by a crew of up to seven and achieves a top speed of 11.7km/h (37mph) on the road. The 152mm gun fires high-explosive fragmentation, submunition and laser-guided projectiles at a maximum rate of eight rounds per minute, and secondary armament consists of a single 12.7mm machine gun. The 2S31 was developed in the mid-1990s and is one of the most advanced Russian weapons currently

deployed. Serviced by a crew of four, the vehicle weighs approximately 17.6 tonnes (19.5 tons) fully loaded. Its maximum rate of fire is 10 rounds per minute, and secondary armament consists of a single 7.62mm machine gun. Modern communications systems combine in the Vena with state-of-the-art protection against nuclear, biological and chemical threats. The vehicle can attain a top road speed of 70km/h (44mph).

The AS-90 155mm self-propelled gun was deployed in the early 1990s with the British Army. Its light armoured protection facilitates speed but adds vulnerability.

Ironically, while the Russian military was being supplied with modern weaponry, shortages of basic materials plagued efforts to defeat the Chechen rebels. Artillery shells have always been in demand.

'The Russian troops in Chechnya have made extensive use of heavy artillery fire to suppress the rebels and this has severely depleted munitions stockpiles, as there has been no serial production of heavy shells in Russia for a decade,' writes Pavel Felgenhauer.

'In the 1994–1996 Chechen war officers complained that they were using shells produced in

the 1980s. In the present conflict shells produced in the 1970s and 1960s were supplied to the front. In December 1999 the Russian government reportedly released 8 billion rubles ($285 million) to buy new heavy shells. But the Russian defence industry has not managed to resume serial production of such munitions.

'Reports from Chechnya say that Russian troops are running out of ammunition for their most used heavy gun – the 122mm D-30 howitzer. One of the remedies being considered in the General Staff in Moscow is to bring out of strategic storage the pre-Second World War M-30 122mm howitzer for which there are millions of rounds, kept since the 1940s.'

A more powerful weapon needed

As the PzH 155-1 SP-70 self-propelled howitzer became obsolete in the mid-1980s, the German, British and Italian governments ended efforts to improve the system in favour of a more powerful weapon. A joint collaboration between the German firms of Rheinmetall and Krauss-Maffei Wegmann eventually produced the Panzerhaubitze 2000, or PzH 2000 (Armoured Howitzer 2000), which is one of the most modern and powerful weapons in service anywhere. The fully loaded vehicle weighs 50 tonnes (55.3 tons), and its main armament of a 155mm L52 howitzer is known for its impressive rate of fire, three rounds in nine seconds during burst mode, 10 rounds in 56 seconds and up to 13 rounds continuously.

While Rheinmetall supplied the main armament for the PzH 2000, Wegmann contributed the chassis, which includes some components of the Leopard 1 main battle tank chassis, and the turret. The PzH 2000 is serviced by a crew of five and

capable of a top speed of 60km/h (37mph) on the road powered by an MTU 881 Ka-500 735kW (986hp) engine. Secondary armament consists of a Rheinmetall MG3 7.62mm machine gun. The PzH 2000 was initially deployed in a combat zone with NATO forces in Afghanistan. In 2006, the Dutch Army fired the PzH 2000 howitzer against Taliban positions in Kandahar province. The British Army has also selected the 155mm howitzer for an upgraded AS-90 called the Braveheart. The PzH 2000 is currently on duty with the armies of Germany, the Netherlands, Greece and Italy.

The French AMX 30 AuF1, which has seen

The Soviet Union's 2S19 Msta self-propelled 152mm howitzer entered service with the Red Army in 1989 and is powered by the chassis of the T-80 main battle tank.

action in the Balkans, incorporates a 155mm GCT gun with the versatile hull of the AMX 30 tank. Serviced by a crew of five, the weapon weighs 39.5 tonnes (43.5 tons) and may fire up to six rounds in 45 seconds. Its top road speed is 60km/h (37.28mph), and the powerplant is a 507kW (680hp) Hispano-Suiza HS-110 engine. A 12.7mm machine gun provides additional firepower. The CAESAR is a recently deployed 155mm self-propelled howitzer, which is currently operated by the armed forces of France and Thailand. Mounted on the wheeled chassis of a Renault six by six truck, the CAESAR is light enough at 16 tonnes (17.7 tons) to be transported by helicopter or aircraft. The design was accepted for production in September 2000, and five were placed in service three years later. The vehicle's top speed on the road is 100km/h (62mph) powered by a diesel engine.

The People's Republic of China has steadily produced new self-propelled artillery pieces during the post-Cold War era. Among the most recent of these are the PLZ45 155mm gun howitzer, the PLZ05 155mm gun howitzer, the SH1 155mm howitzer and the SH2 122mm howitzer. The PLZ45 was developed in the early 1990s specifically for the export market. The 155mm weapon fires a variety of ordnance at a rate of up to five rounds per minute. The vehicle is powered by a 388kW (520hp) turbocharged diesel engine, which allows a top speed of 240km/h (24.85mph). Secondary armament consists of a 12.7mm anti-aircraft machine gun. In 1997 and again in 2001, the government of Kuwait ordered the PLZ45 for its armed forces. The PLZ05 debuted in 2005 at the Beijing International Aviation Expo as an improved variant of the PLZ45. The details on the performance of this weapon are sketchy, although it strongly resembles the Russian 2S19.

The SH2 mounts its 122mm howitzer on a six by six truck, which can travel on the road at nearly 90km/h (56mph). The weapon is probably an adaptation of the PL96 towed cannon, which copied the famed Russian D-30. It can fire up to eight rounds per minute and is serviced by a crew of five. The configuration of the SH122 allows the system to convert from travel to firing mode in as

Short Supply

Russian soldiers carry heavy shells to waiting guns during the fighting in Chechnya. The voracious appetites of Russian guns rapidly depleted available stockpiles, and a crisis developed when the shortage of heavy calibre shells became acute. Russian troops fought two bitter wars in Chechnya beginning in the 1990s in an attempt to suppress rebels fighting to gain independence. The rebels have reportedly aligned themselves with Al Qaeda and other terrorist organizations.

little as 45 seconds. The SH1 155mm howitzer was developed for the export market by China North Industries Group Corporation (NORINCO), and shown to the rest of the world at the 2007 Abu Dhabi International Defence Exhibition. The howitzer is mounted on a six by six truck chassis, and the combined system weighs 20 tonnes (22 tons). Both the SH2 and SH1 utilize sophisticated fire-control systems.

In 1988, the armed forces of the Republic of South Africa fielded the self-propelled 155mm G-6 Rhino gun howitzer. Weighing 43 tonnes (47 tons), the weapon consists of a large turret atop a wheeled chassis. Its configuration is based on the Ratel armored vehicle. Powered by a 525hp diesel engine, the heavy weapon has a top speed of 40km/h (24.85mph) on the road. Its crew of six can operate the main armament at a rate of four rounds per minute, and secondary armament consists of a 12.7mm M2HB machine gun. The Rhino was developed to South African specifications by Canadian Space Research Corporation and served with the South African Army in Angola. Forty of the vehicle are currently in service. Although the Rhino provides excellent firepower and is compatible with NATO ammunition, its extreme weight causes difficulty in long-range transportation.

The Slovak ZUZANA 155mm self-propelled gun howitzer is based on the older DANA 152mm weapons system. It is situated on the chassis of a modified TATRA eight by eight truck, which has a maximum over the road speed of 80km/h (49.71mph). Serviced by a crew of four, its maximum rate of fire is six rounds per minute. A characteristic of the ZUZANA is its gun mounting, which is external between two separate turret

The French-built Caesar 155mm self-propelled howitzer is mounted on a heavy duty Renault truck chassis.

compartments. The weapon is compatible with standard NATO 155mm ammunition.

A way of life

'The people of the Horn have been drowned in a flood of arms from both sides of the Cold War,' wrote Alex de Wall for the *New Internationalist* in December 1992.

'A list of the armaments available in Somalia reads like an inventory of an arms dealer's warehouse. Nine types of artillery piece are in common use on the streets of Mogadishu – five Soviet-made, three US-made and one French

model. The residents of the city refer to the shells they fire as "to whom it may concern" because they are so inaccurate. In addition, two of the heaviest howitzers (one Soviet model and one US) are not currently in use due to lack of trained personnel. Perhaps the most unusual weapon seen on the street is an air-to-air rocket launcher taken from a MIG-21 fighter-bomber and mounted on a Toyota pickup.

'... The quantities sent [by the United States and the Soviet Union] to these two eager clients [Somalia and Ethiopia] was staggering. Ethiopia was equipped with 1400 tanks, over 1000 artillery pieces, 140 fighter-bombers and 35 helicopter gunships. The cost was over eight billion dollars, bought on credit at low interest rates. The last

HEAD TO HEAD: *AS-90* VERSUS

Designed and built by the Vickers subsidiary of BAE Systems, the AS-90 155mm self-propelled cannon was delivered to various units of the British Army in 1993 and has been upgraded from a 39-calibre to a 52-calibre gun. Development of the weapon was begun as a private venture during the 1980s.

AS-90

Crew: 5
Weight: 45,000kg (44.29 tons)
Dimensions: length: 7.2m (23.6ft); width: 3.4m
    (11.2ft); height 3m (9.8ft)
Range: 240km (150 miles)
Armour: 17mm (0.66in) maximum
Armament: one 155mm (6.1in) howitzer
Powerplant: one Cummins 492kW (660hp) V-8 diesel
Performance: maximum road speed: 55km/h
    (34mph); fording: 1.5m (5ft); vertical obstacle:
    0.88m (35in); trench: 2.8m (9.2ft)

STRENGTHS	WEAKNESSES
• Rapid mobility	• Light armoured protection
• Good range	• Needed weapon upgrade
• Design innovation	• Crew vulnerability

PzH2000

Developed and manufactured in Germany by the Krauss-Maffei Wegman and Rheinmetall firms, the Panzerhaubitze 2000, or Tank Howitzer 2000, is armed with a 155mm L52 cannon. One of the most powerful self-propelled artillery weapons currently available, the PzH 2000 was deployed in the 1990s and has served extensively in Afghanistan.

PzH2000

Crew: 3 + 2
Weight: 55,330kg (121,981lb)
Dimensions: length (gun forward): 11.67m (38.3ft);
    length (hull): 7.92m (25.1ft); width: 3.58m
    (11.7ft); height: 3.06m (10ft)
Range: 420km (260 miles)
Armour: 5–17mm (0.197–0.67in)
Armament: one Rheinmetall 155mm (6.1in)
    L52 artillery gun; one MG3 7.62mm (0.3in)
    machine gun
Powerplant: one 736kW (987hp) Ka-500 V8 diesel
Performance: maximum speed: 61km/h (37mph);
    gradient: 50 per cent; vertical obstacle: 1m (3.3ft);
    trench: 3m (9.8ft)

STRENGTHS

- Excellent firepower
- High rate of fire
- Durability

WEAKNESSES

- Limited supply
- Accuracy beyond 40km (25 miles)
- Lack of combat utilization

The South African G6 self-propelled howitzer is shown moving cross-country. This weapon was utilized during the South African Border War of the late 1980s.

rise of global terrorism increases the danger that such weaponry may fall into the hands of those who would strike not only military but also civilian targets. While the world trembled at the prospect of nuclear fallout, it is now experiencing post-Cold War fallout of a different kind – the availability and sometimes indiscriminate use of modern weaponry. Artillery is one of these weapons and has continually been in the news during regional conflicts.

Nevertheless, the modern armed forces of the world continue to require artillery for their own purposes. Towed weapons may be giving way to self-propelled types. However, the basic principle remains the same: to deliver devastating and accurate fire against an enemy, whatever shape it may take on the battlefield.

The modern international standard for the field artillery is the 155mm M777 howitzer, which is manufactured by the British BAE Land Systems company using approximately 70 per cent American-made components. Research and development began with the British-based VSEL in the 1990s. Through a series of mergers and acquisitions, the responsibility for production has been consolidated under BAE, and final assembly is in the United States with an affiliate company. Intended as a replacement for the M198 howitzer with the US Army and the US Marine Corps, it is also projected to supplant the L118 Light Gun with the British armed forces. In 2006, a total of six M777 howitzers were deployed to Afghanistan with the Canadian Army. More than 650 M777s are expected to be deployed with the US armed forces

shipment arrived in March 1991. About one-quarter of Ethiopia's entire gross national product between 1974 and 1991 was spent on the war effort.

'In Somalia, arms expenditure amounted to about ten percent of gross national product; by the end of the decade its army possessed nearly 300 tanks and about 700 artillery pieces and mortars.'

Such is the legacy of the Cold War. The proliferation of arms around the world, sold or handed to proxies battling in the name of political ideology East or West, continues to plague civilization. Countless lives have been lost, and the

by 2010, and Canada has ordered additional numbers as well.

Originally, the M777 was referred to as the Ultralightweight Field Howitzer (UFH). Weighing slightly less than 4082kg (9000lb), the weapon lives up to its former name. The employment of strong but lightweight titanium has contributed to its mobility and allows ease of transportation by helicopter, aircraft or truck. The powerful 155mm howitzer fires at a rate of up to five rounds per minute with a range of up to 30km (18.6 miles) with rocket-assisted ammunition. With an overall length of slightly more than 9.4m (31ft) in towing position, it is stored rather easily as well. The M777 is almost half the weight of the M198, and the required crew of five is reduced from between seven and nine needed to service the M198.

The most advanced field artillery piece in service with the Russian Army is the versatile 120mm Nona-K 2B16 light towed cannon. Mounted on a wheeled carriage with a shield for protection of the crew, the weapon fires at a maximum rate of five high-explosive rounds per minute. The Nona gun has also been mounted on a self-propelled tracked vehicle and on a wheeled amphibious vehicle with the respective designations of 2S9 and 2S23.

Born of the collaboration between various Russian design bureaus, chief among them the Precision Mechanical Engineering Central Research Institute (TsNIITochMash), the Nona incorporates numerous innovative features. These include a compressed-air system fed from the chassis to place rounds in the breech and clear the

barrel after firing; a breech apparatus that allows firing of finned shells and smooth ammunition; and a breechblock that includes a plastic obturator which moves the round into the cannon barrel, locking it into position with its wedge.

Research on the Nona-K was begun in the late 1980s, prompted by the combat experience of Soviet troops in Afghanistan. The weapon is serviced by a crew of five, and the field version is towed by the GAZ-66 or UAZ-469 trucks. One major difference between the towed and self-propelled versions of the Nona-K is that the towed

version has a recoil brake, which absorbs up to 30 per cent of the recoil energy. According to its manufacturers, the Nona-K can fire as a howitzer, mortar or anti-tank gun. Potential customers for the weapon are China, India, Cuba and countries of the former Soviet Union. The Nona-K weighs 13 tonnes (14.5 tons), and its range is 12km (7.46 miles).

GIAT Industries of France produces the LG1, a 105mm towed howitzer that is in service with light infantry battalions of the Canadian Army. The weapon entered service in 1996 and was deployed

Camouflage netting hiding a towed artillery piece. Self-propelled artillery utilizing the shoot-and-scoot philosophy minimizes the risk of detection.

with Canadian units to Bosnia, and has been sold to Belgium, Indonesia, Singapore and Thailand as well. It weight is 1520kg (3351lb), and the 105mm howitzer has a range of 18,500m (20,232yds). The heavier GIAT 155 TR/52 is a towed 155mm cannon weighing 11,000kg (24,251lb) with a range of 41,500m (45,385yds).

Alas Babylon

Bordering on science fiction, the development of the gigantic 'super guns' called Project Babylon seems somewhat appropriate for the delusions of grandeur entertained by Iraqi dictator Saddam Hussein. However, the threat and the potential for destruction these guns presented were very real. At the height of the Iran-Iraq War in the 1980s, Saddam Hussein contacted Gerald V. Bull, a well-known astrophysicist. Bull had previously worked under contract with the United States and Canada and had produced large guns during the 1960s under the auspices of a project called HARP (High Altitude Research Project). He had researched the potential for firing projectiles over great distances with super guns and for launching satellites into orbit. Of course, the prospects for firing nuclear warheads were part of the equation as well.

During HARP, Bull actually conducted research on the Caribbean island of Barbados. His work included firing a projectile from a 5in gun to a remarkable altitude of 70km (43.5 miles). Later, his research moved to a facility in Arizona, where two 16in battleship guns were joined to create a barrel length of 30m (98.4ft). This cannon was capable of

Adopted by both the US Marine Corps and the US Army, the British-designed M777 155mm howitzer is a prime example of the latest in highly mobile field artillery.

firing a specially designed shell called a martlett weighing 84kg (185lb) to an altitude of 180km (111.85 miles). This was first demonstrated in November 1966.

By 1980, Bull was no longer under contract to either the United States or Canada, and he began to expand his efforts with other governments. However, he was tried and convicted of illegally trafficking in arms to South Africa, eventually serving a one-year prison sentence. Reportedly, the Iraqi government spirited Bull to Baghdad aboard a chartered plane, and so began a decade of association between the Canadian artillerist and the Iraqi dictator.

The primary objective of the Iraqi government was the construction of huge artillery pieces. Details are somewhat sketchy, but it appears that during the span of Project Babylon four such weapons were either completed or planned. The first of these was called Baby Babylon and intended for experimentation as a prototype. With a bore of 350mm, the gun's barrel was about 52m (170ft) long. Installed at Jabal Hamrayn about 145km (90 miles) north of Baghdad, Baby Babylon was placed in an excavated position along the side of a mountain at an angle of 45° during the summer of 1989. Its anticipated range was about 668km (415 miles).

Big Babylon, as the second gun was called, was envisioned with a barrel of up to 152m (500ft) in length and a bore fully 1m (3.3ft) wide. The weapon was so ponderously large that it would be virtually impossible to aim it with any degree of precision, and its 1501-tonne (1655-ton) barrel was intended to be supported by a system of cables, harkening back to the construction of the German Paris Gun during World War I. Expanding on his work with HARP, Bull designed a gun that stood

more than 91m (300ft) high with an overall weight of 1905 tonnes (2100 tons). Theoretically, Big Babylon could fire a conventional projectile of 600kg (1322.8lb) more than 998km (620 miles) or a rocket-assisted projectile up to twice that distance. Additional plans included large guns constructed of lighter metal alloys that had been developed during earlier stages of Project Babylon.

Project Babylon lost momentum in the spring of 1990. Bull was assassinated in March, possibly by Israeli Mossad agents attempting to squelch the threat of the super gun, and the following month British customs agents seized eight pieces of the barrel of a second Big Babylon gun. Actually, Baby Babylon was assembled and test-fired. While

Produced by GIAT Industries of France, this LG1 105mm towed artillery piece is being manhandled into a firing position by its crew.

refinements were being made, the first Big Babylon gun was at least partially assembled. However, it was subsequently determined that the supporting framework would never be strong enough to support Big Babylon in fully operational mode.

Following the confiscation of the barrel sections of the second Big Babylon, other parts were seized in Greece and Turkey. Components of the guns had been manufactured in Germany, Switzerland, France, Italy and Spain. In the summer of 1991, the Iraqi government revealed to a United Nations commission that it did indeed possess a gun the size of Baby Babylon, raising questions about the intention of the Iraqis if the guns had become operational during the recently concluded Gulf War.

Commenting on the super gun's development, General Hussein Kamel al-Majeed, an Iraqi defector, noted, 'It was meant for long-range attack

HEAD TO HEAD: *155mm M777* VERSUS

A recently developed 155mm field artillery piece, the M777 howitzer was developed by British Vickers and entered service in 2005 with the US Marine Corps and with the armed forces of Canada. The use of titanium has considerably lightened the weight of the gun.

155mm M777

Calibre: 155mm (6.1in)
Weight: 3745kg (8256lb)
Dimensions (travelling): length: 9.28m (30.4ft);
    width: 2.77m (9.1ft); height: 2.26m (7.4ft)
Elevation: -5° to +70°
Traverse: total 45°
Maximum range: standard projectile: 24690m
    (27,000yds); rocket-assisted projectile: 30,000m
    (32,808yds)

STRENGTHS

• Light weight
• Ease of transport
• Small crew requirement

WEAKNESSES

• Dependence on foreign parts
• Trials and evaluation continuing
• Ammunition availability

Model TR

Manufactured in France by the GIAT firm, the TRF1 155mm field cannon employs combustible casings, which improve the rate of fire due to the fact that no casing must be extracted during operation. The weapon is currently fielded with the armies of France, Saudi Arabia and Cyprus.

Model TR

Calibre: 155mm (6.1in)
Weight (travelling and firing): 10,650kg (23,479lb)
Dimensions: length (travelling): 8.25m (27.1ft); width (travelling): 3.09m (10.1ft); height (firing): 1.65m (5.4ft)
Elevation: +65°/-5°
Traverse: total 65°
Maximum range: standard projectile: 24,000m (26,245yds); rocket-assisted projectile: 30,500m (33,355yds)

STRENGTHS

• Excellent rate of fire
• Rapid deployment
• Varied ammunition

WEAKNESSES

• Heavy weight
• Tube tends to overheat
• Cross-country transport

Employed during the HARP (High Altitude Research Project), undertaken jointly by American and Canadian agencies, a gun fires its projectile at a high trajectory.

and also to blind spy satellites. Our scientists were seriously working on that. It was designed to explode a shell in space that would have sprayed a sticky material on the satellite and blinded it.'

Following the Iraqi admission, UN inspectors located and supervised the destruction of the gun components that were found, including propellant for the shells. Several of the sections of the barrel were seized in Britain in 1990, and are currently on display at the Royal Armouries Fort Nelson, Portsmouth.

The influence of Gerald Bull on the conduct of the Gulf War was felt in other artillery innovations possessed by the Iraqis. In 1985, the Iraqis received 200 of Bull's GC-45 155mm howitzers, which had been manufactured in Austria and designated GH-N-45. Further, the Iraqis possessed the 210mm Al Fao and 155mm Majnoon self-propelled artillery systems. The monstrous Al Fao, weighing 43.5 tonnes (48 tons), was capable of firing a 109kg (240.3lb) projectile 56km (35 miles) at a rate of four rounds per minute. Bull was also instrumental in assisting the Iraqis to modify the warheads of their Scud missiles, increasing their range.

The Gulf War provided an opportunity for many of the state-of-the-art munitions, into which the United States and Coalition forces had poured billions of dollars, to prove their worth, and a number of these systems have been improved during the years since. Anti-aircraft guns of a variety of manufactures were employed by the Iraqis, some directed by radar and others only by visual contact, as evidenced by the haphazard air

defence of Baghdad, images of which were transmitted around the world as their shells lit up the darkened sky. While the guns remain, tactical rockets, anti-tank, and anti-aircraft missile systems continue to fulfill their battlefield roles as well.

Recent weapons

As variants of Cold War and later artillery systems are still in service, several recent anti-aircraft, rocket and anti-tank weapons are worthy of mention. The United States continues to deploy and improve its Patriot anti-aircraft and anti-missile system, while Russia has fielded the 9K37M1-2 Beech, know in NATO nomenclature as the SA-17 Grizzly. The SA-17 continues the improvement of a line of such systems which includes the SA-6 Gainful and the SA-11 Gadfly. The Beech was approved for production in 1995 and reached Russian troops three years later. Mounted atop the 9A30M1-2 self-propelled launcher, it fires two types of missiles, the 9M38M1 or the 9M317. One of the major improvements compared to the SA-11 system is the employment of 9S18M1 'Snow Drift' radar, which tracks a greater number of targets. Each missile weighs approximately 726kg (1600lb), and is capable of attaining altitudes of up to 24,994m (82,000ft). The range of the weapons is approximately 50km (31 miles). A naval version of the Beech has been installed aboard Russian destroyers and aircraft carriers and is called the SA-N-12, and an export variant called the Ural is available. Russian technicians are continually working to improve the 9K38 system.

Although anti-aircraft guns have largely given way to missile systems, the United States has maintained a truck-mounted version of the successful 20mm Phalanx system originally utilized by the US Navy. Among other modern anti-aircraft guns, the CV-9040 Chameleon, of Swedish design, is one of the most interesting. Research into the system began in the early 1990s, and testing had been completed by 1993. Weighing 20.6 tonnes (22.8 tons) and serviced by a crew of three, the Chameleon mounts a 40mm Bofors cannon, which is largely unchanged in mechanics from the original that went into service in 1951. The 40mm gun fires up to 300 rounds per minute and may be particularly effective against aircraft flying low and slow. The weapon may also be used against ground targets.

Utilizing the TDSR-2620 fire-control system, a product of Thompson-CSF of France, the Chameleon aims its weapon with laser range-finding, electronic-optical sights and infrared vision, and ballistic computers. The DS-14 engine is a product of the Swedish SAAB Scania firm and generates 410kW (550hp). Secondary armament consists of a 7.62mm machine gun, and smoke grenade launchers are fitted to the tracked hull.

The Chinese LuDon-200 (Land Shield) is another anti-aircraft system developed from a naval weapon. The LuDon-2000 incorporates the Type 730 30mm weapon system, which houses seven barrels atop the platform of an eight by eight truck. Reportedly, a newer version of the weapon, which includes surface-to-air missiles, is in the works and will be mounted on the Wanshan WS-2400 truck, a Chinese version of the Russian MAZ543 TEL wheeled chassis. Developed for the export market and introduced to the world at the 2005

A partially assembled Babylon gun is shown after its capture during the 1991 Gulf War. The head of the Iraqi Babylon Project, Gerald Bull, was assassinated in 1990.

International Defence Exhibition in Abu Dhabi, the LuDon-2000 is capable of firing up to 5800 30mm rounds per minute. The maximum range is 10,000m (3280yds), and targets are typically taken under fire at ranges of half that distance. Type 347G radar and laser range-finding equipment are operated from an enclosed turret.

The Italian SIDAM 25 anti-aircraft system combines a quadruple mount Swiss 25mm Oerlikon KBA gun atop the US M113 armoured personnel carrier platform. The OTO Breda firm began constructing the SIDAM 25 in 1989 and has taken advantage of cost savings due to the fact that the M113 is in widespread use around the world and parts are readily available at reasonable prices. Powered by a Detroit 6V-53T diesel engine that generates 198kW (266hp), the SIDAM 25 is able to reach speeds of up to 69km/h (43mph) and weighs 15,100kg (33,290lb). The crew of four can produce a rate of fire up to 2400 rounds per minute. Effective against low-flying aircraft, its range is 2027m (6650ft), and a complement of armour-piercing ammunition is usually available for use against enemy ground targets.

HIMARS

The cutting edge of development in rocket artillery is the US High Mobility Artillery Rocket System (HIMARS), which was deployed for the first time in 2005. Development of the HIMARS began in 1996 as part of an Advanced Concept Technology Demonstration program, and production is under the auspices of Lockheed Martin Missiles and Fire Control. In January 2000, Lockheed Martin was asked to produce HIMARS launchers for the US Army, and later two more were manufactured for a US Marine Corps demonstration programme.

On display at Royal Armouries, Fort Nelson, Portsmouth, two sections of a Babylon gun provide some insight as to the size and potential lethality of such a weapon.

In 2002, the Marine Corps reached an agreement with the Army to place 40 of the advanced system in service.

The HIMARS is situated atop one of the American armed forces' newest generations of transport vehicles – the wheeled, six by six, all-wheel drive 4.5 tonne (5 ton) truck, part of the Family of Medium Tactical Vehicles (FMTV). It is also able to fire the total complement of MLRS (Multiple Launch Rocket System) ammunition. The HIMARS, which is comparable to the MLRS M270A1, carrying half the rocket load, is comprised of six rockets or a single Army Tactical Missile System (ATACMS) missile. HIMARS maximizes the concept of shoot and scoot artillery. At 10,886kg (24,000lb), the HIMARS is about half the weight of the MLRS M270 launcher. It can achieve a top speed of 85km/h

(52.82mph) and can acquire a target in as little as 16 seconds. It is light enough for transport by a C-130 aircraft, which is smaller than the C-141 or C-5, therefore allowing access to areas prohibited to the larger planes.

Although the HIMARS has a standard crew of three – a driver, gunner and section chief – it can technically be operated by two soldiers, or even a single soldier, with its computer-based fire-control system, which allows firing orders to be carried out in automatic or manual mode. Current plans are for up to 900 HIMARS launchers to be delivered to the US armed forces. Other nations, such as the United Arab Emirates, have expressed interest in purchasing the system, which is designed for use against light armour, personnel carriers, trucks, artillery and air-defence positions.

During Operation Iraqi Freedom, the invasion of Iraq in 2003, the US Army utilized its latest anti-tank missile system, the FGM-148 Javelin, and achieved spectacular results against the Soviet built T-72 tanks of the Iraqi Army. The Javelin is part of a generation of 'fire-and-forget' missile systems, and its multiple role includes defence against armour and low-flying aircraft such as attack helicopters. Two advantages the Javelin has over most previous systems are the soft launch capability and a top attack method of engaging targets. With the soft launch, the operator may fire the weapon when ready; however, the rocket engine does not fully engage until it has reached a reasonable distance from the launch site. This affords some measure of protection against

The mobile Soviet-made ZRK-SD Kub 3M9, known to NATO as the SA-6 Gainful, was utilized against low-flying enemy aircraft.

backblast, which might injure the operator, and against the telltale signature of the missile, which could bring retaliatory fire. With the soft launch and 'fire-and- forget' combination, the operator may fire and displace immediately, substantially improving survivability. The top attack method brings the missile and its warhead down on top of the target, which is particularly effective against armoured vehicles whose armour protection is thinnest on the top. The weapon can also be fired in direct attack mode.

The Javelin reaches a maximum altitude of 150m (492.1ft) in top attack mode and 50m (164ft) in direct fire mode. Serviced by a crew of two soldiers, a gunner and an ammunition bearer, the Javelin employs the Command Launch Unit (CLU), which targets, fixes and fires the missile. The entire system weighs 22.4kg (49.5lb), and it is reliant on thermal view-finding in target acquisition. Its maximum range is 2500m (2734yds) compared to the BGM-71 TOW missile system, long in service with the US Army, which has a maximum range of 3000m (3280yds). The Javelin impacts its target with tremendous velocity, delivering a shaped charge HEAT warhead of 8.4kg (18.5lb), capable of penetrating armour greater than 51cm (20in) thick.

To date, more than 20,000 Javelin systems and 3000 CLUs have been produced under contract through a joint venture between Raytheon and Lockheed Martin. The weapon may be mounted on the Bradley fighting vehicle in the future; however, its range is a potential issue. The system has been exported to the Republic of China, Great Britain, Ireland, Canada and other countries.

In the summer of 2004, the US government cancelled the development of the LOSAT (Line Of Sight Anti-Tank) surface-to-surface missile. This was intended to destroy tanks and other targets such as helicopters and personnel carriers with an armour-penetrating warhead of solid tungsten, which used kinetic energy to attain substantial velocity. Although the LOSAT programme was terminated, much of the technology obtained during the project is being incorporated into the Compact Kinetic Energy Missile (CKEM). This has been in development since early 2003, was tested in 2006, and successfully destroyed a T-72 tank during a live-fire test sequence at Eglin Air Force Base near Pensacola, Florida. The CKEM weighs 45kg (99.2lb) and has a maximum range of 10,000m (10,936yds). Its kinetic energy penetrator warhead is delivered at a speed in excess of Mach 6.5.

The Russian response to the development of the main battle tanks of the West, such as the US M1A1 Abrams, the Israeli Merkava Mk4 and the German Leopard 2, are the 9M133 Kornet (Cornet) and the 9M123 Khrizantema (Chrysanthemum) anti-tank missiles. Known as the AT-14 Spriggan to NATO, the Kornet was developed by the Russian KBP Design Bureau and debuted in the autumn of 1994. Intended as a replacement for earlier systems such as the AT-4 Spigot and the AT-5 Spandrel, the Kornet is

Swedish Support

The CV9040 is a Swedish infantry fighting support vehicle designed by Hägglunds-Bofors in the early 1990s and manufactured by BAE Systems Hägglunds. The vehicle's main armament is a 40mm Bofors autocannon. The CV9040 can also be equipped with the AMOS advanced mortar system, as well as anti-tank rockets. To date, more than 1000 have been manufactured, and the CV9040 has seen extensive service in Afghanistan.

effective against armour and low-flying helicopters. Its launch weight is 27kg (59.5lb), and its maximum range is 5000m (5468yds). With a SACLOS guidance system, the Kornet warhead is a HEAT shaped charge, which can penetrate armour up to a thickness of about 119cm (47in).

Reports of the Kornet being used in combat during the invasion of Iraq in 2003 have not been verified. However, it is known that the missiles were in the possession of Hezbollah guerrillas during the fighting with the Israeli Defence Force in southern Lebanon in 2006. At the Battle of Wadi Saluki, 11–13 August, Israeli infantry with supporting Merkava IV tanks reportedly engaged Hezbollah fighters armed with the Kornet. Eleven Israeli tanks were damaged, but the defenders were driven off.

The AT-15 is a laser-guided anti-tank missile, which appeared in service with the Russian Army in 2004. It is mounted atop the 9P157-2 tank destroyer, and its radar and laser beam combination HEAT warhead has a maximum range of 6000m (6561yds).

Other modern anti-tank missiles include the fire and forget Israeli Spike, the French ERYX and the Indian Nag. The Spike, which weighs 40kg (88lb) including the launcher, is capable of delivering a shaped HEAT warhead nearly 8047m (8800yds). The fourth generation of anti-tank missiles developed by the Rafael company, the Spike is intended to replace the Dragon and MILAN systems. The ERYX was developed in the late 1980s and was initially deployed in 1994 under a joint venture between the French and Canadian governments. The shoulder fired ERYX weighs only 10.2kg (22.5lb)and has an effective range of up to 600m (656yds) with its SACLOS-

guided HEAT warhead of 3.6kg (7.9lb). Ideal for urban warfare, the ERYX may be fired from confined spaces such as a small room or alley. The Nag, developed by India's Integrated Guided Missile Development Programme (IGMDP) which began in 1983, was test-fired successfully several times during the late 1990s and up to 2002. Powered by a solid fuel rocket, the infrared-guided Nag has a range of nearly 6035m (6600yds).

A battery of M270 MLRS self-propelled loader/launchers fires a near simultaneous barrage of missiles. The chassis of the M270 is a variant of the Bradley Fighting Vehicle.

Target: Air Force One

The proliferation of arms during the post-Cold War era has fuelled a thriving black market in weapons. As a result of this covert arms trafficking, state-of-the-art munitions have, at times, fallen into

HEAD TO HEAD: *Javelin light forces antitank guided weapon* VERSUS

The Javelin missile system was developed in the late 1980s for the U.S. Army and US Marine Corps. The missile was test fired in 1993 and deployed to infantry units two years later. The weapon replaced the outdated Dragon system.

Javelin

Length: 1.08m (3.5ft)
Diameter: 126mm (5in)
Missile weight: 11.8kg (26lb)
Total weight: 22.3kg (59lb)
Guidance: Lock-on begore launch, automatic self-guidance
Warhead: tandem-shaped charge
Prolulsion: 2-stage solid propellant
Range: 2500m (8202ft)

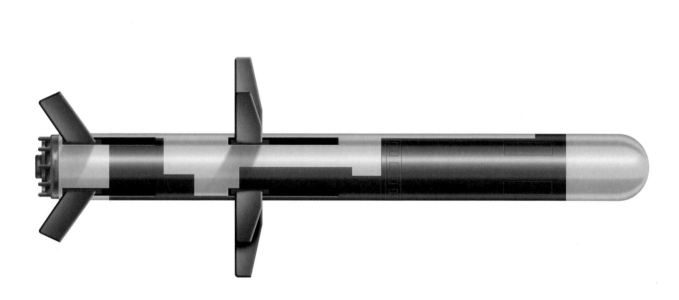

STRENGTHS

- Fire-and-forget technology
- Medium anti-tank capability
- Infantry portable

WEAKNESSES

- Firing exhaust plume
- Targeting requires heat signature
- Cost limits live-fire training

Spike Anti-Armour Missile System

The Spike missile system is the latest such anti-tank system produced by the Israeli manufacturing firm of Rafael. Powered by a solid fuel rocket engine, the missile was placed in service in 1997 and has been exported to at least a dozen other countries.

Spike

Seeker: CCD, IIR or dual CCD/IIR
Length (Spike-MR/Spike-LR): 1200mm (47in)
Diameter (Spike-MR/Spike-LR): 130mm (5.1in)
Range (Spike-MR): 200–2500m (656–8202ft)
Range (Spike-LR): 200–4000m (656–13,123ft)
Weight (missile in canister): 13kg (8818lb)

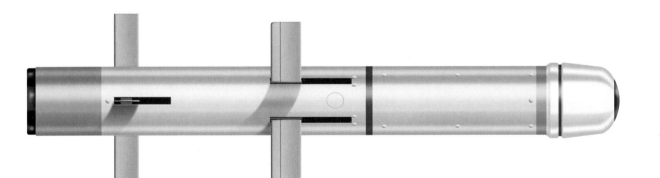

STRENGTHS

- Fire-and-forget technology
- Soft launch capability
- Infantry portable

WEAKNESSES

- Firing exhaust plume
- Limited availability
- Development competition

the wrong hands. Such was the case in the summer of 2003 when a British citizen, Hemant Lakhani, was arrested while attempting to bring a Russian-made, Igla, shoulder-fired, anti-aircraft missile into the United States. Captured as a result of cooperation between US, British, and Russian intelligence agencies, Lakhani thought he was smuggling an operational Igla into the country. However, Russian agents sold a non-working missile to him, and US agents posed as terrorists wishing to buy the missiles in quantity. Lakhani was arrested in Newark, New Jersey, attempting to deliver the Igla to the undercover US agents. He had reportedly intended to target Air Force One, the aircraft that transports the president of the United States. He had further intended to purchase up to 50 more of the missiles.

The artillery of tomorrow

In a presentation made in the spring of 2000, J. Frank Thompson, vice-president of engineering design and development for US defence contractor General Dynamics, argued that several combat deficiencies remain with modern artillery. Even considering rapid deployment, shoot-and-scoot mobility, elaborate guidance systems and smart weapons, Thompson observed that modern artillery is strategically non-deployable. Among his observations were the fact that a brigade may require as much as 96 hours, a division 120 hours, and five divisions as long as 30 days. Further, he noted that there is a lack of small, survivable and lightweight platforms, and the systems themselves lack sufficient lethality, are unable to fight upon arrival, and are costly to sustain.

Such issues present opportunities for improvement, which are being addressed via

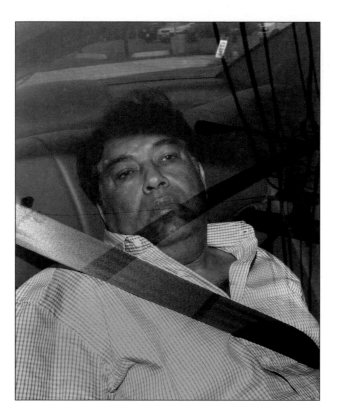

British arms dealer Hemant Lakhani arrives at court in New Jersey to answer charges that he attempted to smuggle a Russian shoulder-fired anti-aircraft missile into the US.

initiatives undertaken by entities such as the US Defense Advanced Research Projects Agency (DARPA) and its Future Combat Systems (FCS) programme. FCS is intended to serve as the foundation for the US armed forces of the future, integrating all branches of service. The US Army programme office is heading the project with Boeing and Science Applications International Corporation as systems integration partners. FCS will link platforms, weapons and detection systems together

and is scheduled to continue testing, which has been successful and was well underway in 2004. However, budget cuts have impeded the progress of the $340 billion project although its testing and development are expected to continue beyond 2010.

One example of a future artillery system is the BAE Systems Land and Armaments Non-Line-of-Sight Cannon (NLOS). A component of the FCS initiative, the NLOS cannon utilizes technology from the cancelled XM2001 Crusader project and is intended to provide heavy fire in support of Combined Arms Battalions (CABs). The NLOS combines rapid mobility and effective fire utilizing a variety of ordnance. Relatively lightweight, it employs a fully automated firing system.

The procurement of advanced weapons systems by the British Army during the next decade is expected to cost up to £2.5 billion. Among the enhancements being considered currently is the Extended Range Ordnance/Modular Charge System (ERO/MCS), which is intended to extend the range of the AS-90 155mm self-propelled howitzer from 24,700m (27,012yds) to 30,000m (32,808yds) when firing high-explosive conventional shells. ERO/MCS includes the fitting of a larger 52-calibre barrel on the howitzer rather than the original 39-calibre, and a two-part charge that is an improvement over the older L-series charge.

In the spring of 2006, Christopher F. Foss, Chairman of the 2006 Future Artillery Conference, invited colleagues to attend the gathering. His invitation eloquently spoke of the days ahead:

'There has never been a greater need than there is now for indirect fire precision strike capabilities…The increased emphasis being placed on rapid deployment forces is leading to the development and fielding of much lighter artillery systems. In some cases, these have similar capability to existing systems, while others are deficient in range and capabilities. Future requirements demand that the field artillery must have a precision strike capability. But at the same time, it must retain its suppressive fire mission. Modern artillery systems are being asked to do more.'

An artillery alternative?

Considering the amount of effort and capital expended to improve artillery during the post-Cold War era, it is somewhat surprising that a school of thought does exist which might herald the end of artillery as we know it. During the spring of 2006, Lutz Unterseher described the potential for replacement of artillery by smaller mortars. He told the Project on Defence Alternatives:

'Somewhat before the demise of the Crusader system, which would have been the world's heaviest mechanized howitzer, a vivid debate on future artillery began. This has further intensified in the related debate between the proponents of solid armour and the advocates of "travelling light".

'Some military experts believe that artillery, particularly its mechanized variant, has lost much ground to the relatively simple and rugged mortar for indirect-fire support…Mechanized artillery systems, which give protection to their crews, are said to be far too heavy to meet the challenge posed by contemporary scenarios that require speedy operational or strategic deployment. In contrast, mortars, weighing only several hundred kilos, can be used in paradrop operations and also issued to heliborne infantry…
'The proposed alternatives to modern

artillery are not as convincing, however, as they appear at first glance… Such weapons [as the mortar] have quite limited effective ranges… Likewise there is a significant limitation on their calibre. In general it is not larger than 120mm. Mortars with larger calibres have been phased out nearly everywhere. Their clumsiness, high weight and forceful recoil, which demand fairly heavy, tracked platforms, neutralize the key advantages of the mortar, namely its lightness and flexibility. By comparison, standard tube artillery has a larger calibre

(West: 155mm/East: 152mm) than practically all mortars. As a result there is much more volume for explosives and warhead sophistication – i.e. bomblets and precision guidance.'

It seems, therefore, given the past, present and future conduct of warfare, that there will always be a place for well-commanded artillery on the battlefield.

The Non-Line of Sight Cannon (NLOS-C), an advanced 155mm howitzer, is designed for rapid deployment, extreme battlefield mobility and decisive firepower.

Glossary

AA Anti-aircraft.

AFV Armoured fighting vehicle.

AP Armour piercing.

APCR Armoured piercing cored round, ammunition with a hard core (usually tungsten).

APDS Armoured piercing discarding sabot.

Battery Descriptive term for when a cartridge is in place and the gun is ready for firing.

Bolt The part of a firearm which usually contains the firing pin or striker and which closes the breech ready for firing.

Blowback Operating system in which the bolt is not locked to the breech, thus it is consequently pushed back by breech pressure on firing and cycles the gun.

Breech The closed end of the barrel (also used to measure its length.

Breech-block Another method of closing the breech which generally involves a substantial rectangular block rather than a cylindrical bolt.

Bullpup Term for when the receiver of a gun is actually set in the butt behind the trigger group, thus allowing for a full length barrel.

Carbine A shortened rifle for specific assault roles.

Calibre The inside diameter of the barrel (also used to measure its length).

Chamber The section at the end of the barrel which receives and seats the cartridge ready for firing.

Closed Bolt A mechanical system in which the bolt is closed up to the cartridge before the trigger is pulled. This allows greater stability through reducing the forward motion of parts on firing.

Compensator A muzzle attachment which controls the direction of gas expanding from the weapon and thus resists muzzle climb or swing during automatic fire.

Delayed Blowback A delay mechanically imposed on a blowback system to allow pressures in the breech to drop to safe levels before breech opening.

Double action Relates to pistols which can be fired both by cocking the hammer and then pulling the trigger, and by a single long pull on the trigger which performs both cocking and firing actions.

Effective Ceiling The highest altitide (of an approaching aircraft) to which a AA gun can fire for 30 seconds before it reaches maximum elevation.

Effective Range The furthest distance a weapon can be accurately aimed.

Elevation The amount a gun can be moved vertically.

Flash Suppressor A device that minimises the visible flash from the gun when fired.

Flechette An bolt-like projectile which is smaller than the gun's calibre and requires a sabot to fit it to the barrel. Achieves very high velocities.

Gas Operation Operating system in which a gun is cycled by gas being bled off from the barrel and used against a piston or the bolt to drive the bolt backwards and cycle the gun for the next round.

GPMG General Purpose Machine Gun.

HE High exlplosive.

HESH High explosive shaped head, ammunition with a shaped charge warhead.

LMG Light Machine Gun.

Locking The various methods by which the bolt or breech block is locked behind the chamber ready for firing.

Long Recoil A method of recoil operation where the barrel and bolt recoil for a length greater than that of the entire cartridge, during which extraction and loading are performed.

MG Machine gun.

Muzzle The front, open end of the barrel

Muzzle Brake A muzzle attachment which diverts muzzle blast sideways and thus reduces overall recoil.

Open Bolt A mechanical system in which the bolt is kept at a distance from the cartridge before the trigger is pulled. This allows for better cooling of the weapon between shots.

PDW Personal Defence Weapon. A compact firearm, smaller than a regular assault rifle but more powerful than a pistol, intended as a defensive weapon for personnel whose duties do not normally include small arms combat.

Penetration (of armour) Given in the form AA/BBB/C, where AA is the thickness of armour penetrated in millimetres; BBB is the range at which it occurred; and C is the slope of the armour. Thus, 75/1000/30 degrees means that the shot penetrated 75mm (2.95-in) of armour at 1000 metres range, striking at an angle of 30 degrees to the target face.

Receiver The body of the weapon which contains the gun's main operating parts.

Recoil The rearward force generated by the explosive power of a projectile being fired.

Recoil Operated Operating system in which the gun is cycled by the recoil-propelled force of both barrel and bolt when the weapon is fired. Both components recoil together for a certain distance before the barrel stops and the bolt continues backwards to perform reloading and rechambering.

Rifled (barrel) A barrel with spiral grooves that make the shell spin, for greater accuracy.

Sabot A protective sleeve that fits round a (usually finned) shell fired from a smoothbore gun.

SAW Squad Automatic Weapon.

Self-loading Operating system in which one pull of the trigger allows the gun to fires and reload in a single action.

Shaped (charge) Explosive that is shaped or becomes shaped on impact in a way that gives it maximum destructive value when it burns.

Short Recoil A compressed version of recoil operation in which the barrel and bolt move back less than the length of the cartridge before the bolt detaches.

Shrapnel Ammunition that, when it explodes, spreads small pieces of hot metal in all directions, most effective against infantry.

Smoothbore A barrel that does not have rifled grooves.

Traverse The amount a barrel can be moved horizontally.

Index

Page numbers in *italic* refer to illustrations
Page numbers in **bold** refer to Head to Head comparisons
Service ranks are the latest shown in the text

(**AA**) = anti-aircraft gun; (**ATG**) = anti-tank gun;
(**ATM**) = anti-tank missile; (**FG**) = field gun;
(**GH**) = gun howitzer; (**H**) = howitzer;
(**RG**) = railway gun; (**RW**) = recoilless weapon;
(**SP**) = self-propelled gun

Afghanistan 184, 199, 204
American Civil War 11–17
    see also under United States guns
Antietam, Battle of 14–15
Australia: 1st Australian Division 39
Austria-Hungary
    Austrian 21st Division 56
    World War I guns
        149mm Skoda 14 (**H**) **51**
        380mm Skoda (**RG**) 60
        420mm Skoda (**RG**) 60

Balaclava, Battle of 10–11
Boer War 34, 35, 37
Brazil: Astros II MLRS 173, *174*
British Army
    BEF 34, 35, 36, *37*, 67, 105, 122
        II Corps 35, 36
        11 Battery 35, 36
        37 Battery 37
        122 Battery 35, 37
        123 Battery 37
        124 Battery 37
    divisions
        7th Armoured 90
        Eighth Army 70
    regiments
        2nd Manchester 36
        7th Hussars 90
        11th Hussars 90
        Argyll and Sutherland Highlanders 36
        King's Own Royal Lancasters 36

King's Own Yorkshire Light Infantry 35
        Parachute 113–14
        Royal Artillery 117, 120, 134
        Royal Field Artillery 34
        Royal Garrison Artillery 37, 38
        Royal Horse Artillery 34, 70
        Royal Scots 36
        Royal Scots Fusiliers 39
        Suffolks 35, 36
        West Kents 37
British Commonwealth
    20mm Polsten (**AA**) 108
    40mm Bofors (**AA**) 104
British guns
    Crimean War
        12-pdr Armstrong 9–10
        cannon *11*
    World War I
        3.3in QF 18-pdr (**FG**) 27, *28*, *42*
        4.5in QF (**H**) 34, 35, *50*
        4.5in QF (**H**) Mk II 35
        4.7in QF (**FG**) 35, *36*
        5in 60-pdr 37–8
        6in (**H**) 38
        8in (**H**) 35
        9.2in Mk I (**H**) 35, 38–9
        9.2in Mk II (**H**) 39
        12in Mark I (**RG**) 39, 60
        12in Mark II (**H**) 39
        12in Mark III (**RG**) 39
        12in Mark IV (**H**) 39
        13-pdr QF (76mm) 34, *35*
        14in (**RG**) 60
        15in (**H**) 35
        15-pdr (84mm) 35
        18-pdr QF 34
        18-pdr QF Mk II 34
        18-pdr QF Mk IV 34
        60-pdr BL Mk 1 *39*
    World War II
        2-pdr QF (**ATG**) 117
        2-pdr QF Mark III (**ATG**) 117
        2-pdr QF Mark VII (**ATG**) *117*

3in QF 20cwt (**AA**) 104–5, *104*
        3.7in QF Mark I (**AA**) 105, 108, *108*
        3.7in QF Mark II (**AA**) 105, 108
        3.7in QF Mark III (**AA**) 105, 108
        4.5in QF Mk II (**AA**) 108
        6-pdr QF (**ATG**) 117
        6-pdr QF Mark II (**ATG**) 117, **119**
        6-pdr QF Mark IV (**ATG**) 117
        7.2in M I-V (**H**) 71
        7.2in M6 (**H**) 71
        17-pdr QF (**ATG**) 117, 120, *121*
        25-pdr Bishop (**SP**) *125*, 126
        25-pdr QF Mk I 67
        25-pdr QF Mk II 67, **82**
        25-pdr Sexton (**SP**) 126
    Cold War
        105mm FV433 Abbot (**SP**) 157, *160*
        105mm L118 **142**, 146
        105mm L119 146
        105mm Mod 56 (**H**) 146
        120mm L6 Wombat (**RW**) 163–4, *165*
        155mm AS-90 (**SP**) 157
        155mm FH-70 146–7, **149**
    modern artillery
        155mm AS-90 (**SP**) 198, *198*, **202**
        155mm M777 (**H**) 204–5, *206*, **208**
        155mm NLOS (**SP**) 218, *219*
British missiles
    Bloodhound 180, *180*
    Malkara/Vigilant 166–7, *167*
    Rapier 181–2, *182*
    Thunderbird/II 181, *181*
British rockets
    2in rocket 131
    3in rocket *130*, 131
    Land Mattress 132
    LILO 131–2, *131*
Bull, Gerald 206–7, 210
    155mm GC-45 (**H**) 147
Burney, Denis 163

Canada
    105mm LG1 (**H**) 205–6

155mm M777 (**H**) 204–5
ERYX missile 215
HOT 1 (**ATM**) 168
Chechnya 198–9, *200*
Chinese guns
    Cold War
        100mm Type 73 (**ATG**) 164
        100mm Type 86 (**ATG**) 164
        105mm Type 75 (**RW**) 164
        152mm D-20 147
        152mm Type 83 147, 157, 160
        155mm PLL01 (**GH**) 147
        155mm Type 89 147, *150*
    modern artillery
        30mm LuDon-2000 (**SP**) 211–12
        122mm SH2 (**SP**) 200, 201
        155mm PLZ05 (**SP**) 200
        155mm PLZ45 (**SP**) 193, 200
        155mm SH1 (**SP**) 200–1
Chinese missiles
    Kai Shan 1 (**AA**) 185
Chinese rockets
    107mm Type 63/81 173
Churchill, Sir Winston 39, 141, *144*
Crimean War 9, 10–11, 125
Czechoslovakian guns
    interwar period
        47mm Skoda PUV vz 36 (**ATG**) 117
        76.5mm Skoda vz 30 87–8
        100mm Skoda vz 30 (**H**) 87–8
        149mm Skoda vz 37 (K4) (**H**) 86–7
    Cold War
        152mm DANA ShKH77 (**SP**) 156–7, *157*
        155mm ShKH Zuzana (**SP**) 157

Dien Bien Phu 147, 150, *151*
Dunkirk 34, 67, 105

early artillery
    pre-Industrial Revolution 7–8
    start of modern era 8–9
Egypt: 180mm S-23 gun 146
El Alamein 70, 80, 126

Falklands War 182, 184
Franco-Prussian War 17–20, 28
French guns
  Franco-Prussian war
    13mm Mitrailleuse 1866 18
    Chassepot rifle 18
  World War I
    58T trench mortar 32
    75mm 1897 (**FG**) 28–9, *29*, **30**, 52
    75mm Schneider 1912 32
    105mm Schneider 1913 **44**
    155mm Rimalho (**H**) *29*, 32–3, *33*
    155mm Schneider (**H**) 32
    520mm Schneider Obusiers (**RG**) 60–1
  interwar period
    105mm Schneider M1934 S 72
    105mm Schneider M1935 B 72
  World War II
    25mm SA-L M1934 122
    25mm SA-L M1937 122
    47mm SA M1937 (**ATG**) 121
    47mm SA M1939 121, *122*
    75mm 1897 (**FG**) 71
    105mm Schneider M1935 B (**FG**) *72*
  Cold War
    75mm (**RW**) 161–2
    105/155mm AMX–13 (**SP**) 160, *161*
    155mm AUFI (**H**) 147
    155mm TR (**H**) 147
  modern artillery
    105mm LG1 (**H**) 205–6, *207*
    155mm AMX 30 AuF1 (**SP**) 199
    155mm CAESAR (**SP**) 199–200, *201*
    155mm TRF1 (**FC**) **209**
  French missiles/rockets
    227mm LRM rocket 173
    Entac 167
    ERYX (**ATM**) 215
    Hades 194
    HOT 1 (**ATM**) 168
    Pluton 194
    Roland (**AA**) 184, *185*

German Army
  15th Panzer Regiment 80
  Afrika Korps 67, 70

armies
  Sixth Army 74
  Ninth Army 127, 130
  Motorized Artillery 84, 85
German guns
  World War I
    77mm Feldkanone 16 52
    77mm Feldkanone M96nA **31**, 52
    88mm M1913 SK L/45 53
    100mm M04 52
    100mm M14 52–3
    100mm M17 53
    105mm 98/09 (**H**) 53–4, *54*
    150mm sIG 33 **45**
    170mm Schnelladekanone 53, 60
    210mm Paris Gun 59–60, *60*
    250mm *minenwerfer* 54
    280mm (**H**) 46
    305mm 1911 mortar 46, *53*, 55, *56*–7
    380mm Langer Max (**RG**) 55, 60
    420mm L/14 (**H**) ('Big Bertha') 46, **48**, 55
  interwar period
    37mm Pak 35/36 (**ATG**) 116
    75mm FK 16 nA 86
    75mm leFK 18 86
    88mm Flak 18 (**AA**) 79, *79*
    88mm Flak 36 (**AA**) 79
    88mm Flak 37 (**AA**) 79
    105mm Cannon 18 85
    105mm leFH 18 85
    150mm (**H**) 18 81, 84
    150mm 37(t) (**H**) 87
    220mm (**H**) 87
    240mm Cannon 3 85
    355mm M1 (**H**) 84
  World War II
    20mm Flak 28/29 (**AA**) 108
    20mm Flak 30 (**AA**) 103
    20mm Flak 38 (**AA**) 103, *103*
    28mm sPzB 41 (**ATG**) 116
    37mm Flak 18 (**AA**) 103
    37mm Flak 36 (**AA**) 103
    37mm Flak 37 (**AA**) 103
    37mm Flak 43 (**AA**) 103, 104, **107**
    40mm Flak 28 (**AA**) 104
    42mm lePak 41 (**ATG**) 116–17
    47mm Pak 36(t) (**ATG**) 117

    47mm Pak 141(f) 121
    50mm Flak 41 (**AA**) 104, *105*
    50mm Pak 38 (**ATG**) 116
    75mm 1897 (**FG**) 71
    75mm Cannon 75/32 Model 37 89
    75mm Flak Vickers(e) (**AA**) 105
    75mm Pak 40 (**ATG**) 116, **118**
    75mm Pak 41 (**ATG**) *116*, 117
    75mm Pak 97/38 71
    75/105mm StuG III/G (**SP**) 123
    75mm StuG IV (**SP**) 123
    76.2mm FK 296(r) 74
    76.2mm Pak 36(r) (**ATG**) 74, 121
    77mm Model 16 (**FG**) *87*
    88mm Flak 18 (**AA**) 80–1, *80*
    88mm Flak 36 (**AA**) 80–1
    88mm Flak 37 (**AA**) 80–1
    88mm Flak 41 (**AA**) 79–80, **81**
    88mm Pak 43/41 (**AA**) **83**
    94mm Flak Vickers M39(e) (**AA**) 108
    100mm Skoda vz 14/19 (**H**) 88
    105mm Cannon 18/40 85–6
    105mm Flak 38 (**AA**) 102, *102*
    105mm Flak 39 (**AA**) 102
    105mm Flak 40 (**AA**) 102
    105mm leFH 18 (**H**) **77**
    105mm leFH 18/40 (**H**) 85
    105mm leFH 18(M) (**H**) *84*, 85
    105mm Schneider 1913 71
    105mm Waffentrager (**SP**) 126
    105mm Wespe (**SP**) 122, *123*, **128**
    128mm Gerät 40 (**AA**) 102
    149mm Skoda 18 (**H**)
    150mm (**H**) 18 84, *86*, 87
    150mm SdKfz 138/1 (**SP**) 123
    150mm sFH18 (**H**) **69**
    150mm sIG 33 (**SP**) 123
    150mm sIG 38(t) (**SP**) 123
    150mm Sturmpanzer IV Brummbär (**SP**)
     123, 125
    170mm Kanone 18 84
    210mm Howitzer 520(i) 90
    210mm K12 (E) (**RG**) *138*
    210mm M^rser 18 84
    220mm Skoda (**H**) *89*
    283mm K5 (**RG**) 138–9
    355mm M1 (**H**) 84

    380mm Sturmtiger (**SP**) *124*, 125–6
    520mm Langer Gustav 139
    540/600mm Karl series (**SP**) 125
    800mm Schwerer Gustav/Dora (**RG**) 139,
     139
    V-3 cannon 139
  modern artillery
    155mm PzH 2000 (**SP**) 199, **203**
German missiles
  HOT 1 (**ATM**) 168
  Roland (**AA**) 184, *185*
  Wasserfall (**AA**) 178
German rockets
  28/32cm Wurfk^rper sdKfz 251 **137**
  150mm Wurfgranate 41 133, *133*
  210mm Wurfgranate 42 133
  280mm Wurfk^rper 133
  300mm Wurfk^rper 42 133, *134*
  320mm Wurfk^rper 133
German States
  4-pdr (77mm) C/64/67 19
  6-pdr (91.6mm) C/61 19
Germany, Federal Republic of
  155mm FH-70 (**H**) 146–7, **149**
Gettysburg, Battle of *14*, *16*, 17
Greece: 155mm PzH 2000 (**SP**) 199
Gulf War, First 188–9, 190, 194–5, 196, 210
  Coalition guns
    155mm M109A2 (**SP**) 195
    155mm M109A6 Paladin (**SP**) 196, *196*
    155mm M198 (**H**) 196

Hitler, Adolf 48, 74, *78*, 79, 81, 86, *89*, 113
Hungary
  80mm Bofors (**AA**) 104
  149mm Skoda Model 14 (**H**) **51**
  210mm Obice 210/22 Model 35 (**H**) 90
Hussein, Saddam 188, 206

India
  NAG (**ATM**) 215
  Pinaka rocket launcher 173
Iran–Iraq War 196, 206
Iraq
  155mm GH-N-45 (**H**) 210
  155mm Majnoon (**SP**) 210
  210mm Al Fao (**SP**) 210

Baby Babylon 207, *211*
Big Babylon 207, 210, *212*
Scud missile 188–90, *191*
Israeli guns
 Cold War
  155mm L-33 Ro'em (**SP**) 161, *162*
  155mm M109 (**SP**) 161
  160mm Makmat 160 (**SP**) 161
  175mm M107 (**SP**) 154
Israeli missiles
 Spike (**ATM**) 215, **217**
Italian guns
 interwar period
  75mm Obice 75/18 Model 35 (**H**) 89–90, *90*
 World War II
  20mm Breda (**AA**) 111
  20mm Cannone-Mitraglia da 20 Oerlikon (**AA**) 109
  20mm Scotti (**AA**) 111
  75mm Cannon 75/32 Model 37 89
  90mm Cannon 90/53 (**AA**) 111
  210mm Obice 210/22 Model 35 (**H**) 90
 Cold War
  105mm Mod 56 **143**, 147
  155mm FH-70 (**H**) 146–7, **149**
 modern artillery
  25mm SIDAM 25 (**SP**) 212
  155mm PzH 2000 (**SP**) 199

Japanese guns
 Russo-Japanese war
  4.7in cannon 21
  11in (**H**) 22, 23
  12-pdr cannon 22
 interwar period
  37mm Type 94 121
  37mm Type 97 121
 World War II
  20mm Type 98 (**AA**) 111
  37mm Type 97 (**ATG**) 116
  47mm Type 1 (**ATG**) 121
  75mm Type 2 (**SP**) 127
  75mm Type 35 gun **95**
  75mm Type 38 (Improved) (**FG**) 93
  75mm Type 88 (**AA**) 110–11
  150mm Type 4 HO-RO (**SP**) 127

Japanese rockets
 200mm rocket 133
 447mm rocket 133

Korean War 144, 153
Krupp 10, 19, 22, 46–8, *49*, 52, 79, 81

Maginot Line 72–3, *73*, 121, 125, 139
Manchuria 93
Mons 34, 35
Montgomery, Gen. Bernard 70
Mussolini, Benito 88, *89*

NATO 141, 146–7, 150–1
 M270/A1 MLRS rocket 173
 P155mm zH 2000 (**SP**) 199
Netherlands: 155mm PzH 2000 (**SP**) 199
Neuve Chapelle, Battle of 38
New Zealand: 5th Field Regiment 132–3
North Vietnam: SA-2 Guideline 175, *175*

Pétain, Gen. Philippe 55
Poland: Skoda 220mm M28 (**H**) 87
Port Arthur, siege of 21–3, *24*, *25*, 91
Prussia: 1000-pdr cannon 20

Rheinmetall 52, 79, 81
Romania: Skoda 149mm vz 37 (K1) (**H**) 86
Rommel, Field Marshal Erwin 67, 79, 80–1
Roosevelt, Franklin D. 99, 110
Russian Army
 brigades
  5th East Siberian Artillery 23, 24
  35th Artillery 23
  5th East Siberian Rifle Division 24
Russian Artillery Corps 57
Russian Civil War 58
Russian guns
 World War I
  91mm M16 trench mortar 57
  122mm M09 (**H**) 57
  122mm M10 (**H**) 57
  Putilov 76.2mm M02 57–8
Russo-Japanese War 20–4, 57, 90, 91
Russo-Turkish War 10

Sedan, Battle of 20, *21*

Sino-Japanese War 92
Six-Day War 161
Skoda 86–8
Slovakia: 155mm ShKH ZUZANA (**SP**) 201
Somalia 201, 204
South Africa: 155mm G-6 Rhino (**SP**) **192**, 201, *204*
Soviet guns
 interwar period
  76.2mm M00/02 (**FG**) 73
  76.2mm M02/30 (**FG**) 73
  76.2mm M1936 (**FG**) 73–4
  152mm (**H**) 74
  203mm M1931 (**H**) 74
 World War II
  37mm Pak 35/36 (**ATG**) 121
  76.2mm M1939 (**ATG**) 121
  76.2mm M1942 (**FG**) 74, **76**, 121
  76.2mm SU-76 (**SP**) 127
  85mm Cannon (**AA**) 110
  122mm ISU-122 (**SP**) 127, *127*
  152mm ISU-152 (**SP**) *126*, 127
  152mm M1910/34 (**H**) 74
  152mm M1937 (**H**) 74, *74*
 Cold War
  73mm SPG-9 (**RW**) 163
  76.2mm M1969 (**ATG**) 146
  82mm B-10 (**RW**) 162–3
  100mm 2A19 (T-12) (**RW**) 163, *164*
  100mm 2A29/MT-12 (**RW**) 163
  122mm 2S1 Gvozdika (**SP**) 155–6
  122mm D-74 145
  122mm D30 (**H**) 145, **148**
  125mm 2A45 Sprut B (**RW**) 163
  130mm M-46 (**FG**) *147*
  152mm 2S3 (**SP**) **159**
  152mm D-1 146
  152mm D-20 (**H**) 146, *146*
  152mm M1973 (2S3) (**SP**) 156
  152mm M1976 146
  152mm M1981 (2S5) (**SP**) 156, *156*
  152mm M1987 146
  180mm S-23 146
  240mm M1975 (2S4) (**SP**) 156
 modern artillery
  120mm 2S31 **SP** Vena 198
  120mm Nona-K 2B16 205

122mm D-30 **H** 199
152mm 2S19 MSTA-S **SP** 198, *199*
Soviet missiles
 9K714B Tochka 191
 AT-1 Snapper 167
 AT-3 Sagger (**ATM**) 167, **171**
 At-7 Saxhorn (**ATM**) 167
 AT-14 Spriggan (**ATM**) 214–15
 AT-15 (**ATM**) 215
 FROG variants 172, *172*
 PR-90 190
 R-17 Scud 190
 SA-1 Guild (**AA**) 178–9
 SA-2 Guideline (**AA**) 179
 SA-3 Goa (**AA**) 179, *179*
 SA-6 Gainful (**AA**) 210, *213*
 SA-10 Grumble (**AA**) 179–80
 SA-11 Gadfly 210
 SA-16 Gimlet (**AA**) 180
 Sa-17 Grizzly 210
 SA-18 Grouse (**AA**) 180
 SA-19 Grisom (**AA**) 179
 SS-21 Scarab 172, *173*
 SS-23 Spider 190–1, *195*
 SS-26 Iskander 191
Soviet rockets
 82mm M-8 130
 122mm BM-21 172, **176**, 187
 132mm M-13 127, 130, *132*, **136**
 132mm M-13-DD 130
 140mm BM-14 172
 220mm BM-27 172–3, *189*
 240mm BM-24 172
 300mm BM-30 173
 300mm M-30 131
 300mm M-31 131
Spanish Civil War 79, 116
Spicheren, Battle of 20, *20*
Stalin, Joseph 73, 74, 99
Stalingrad 127
Stones River, Battle of *6*, 15
Swedish guns
 World War II
  40mm Bofors L/60 (**AA**) 104, **106**
  75mm Bofors (**AA**) 104
  80mm Bofors (**AA**) 104

Cold War
  Carl Gustav (**RW**) M1/M2/M3
    162, *163*
modern artillery
  40mm CV-9040 Chameleon (**AA**) 211, *214*
Switzerland
  20mm Oerlikon (**AA**) 108, *109*
  Cobra missile 167

Thailand: 155mm CAESAR (**SP**) 199–200
Tobruk 67, 70
Truman, Harry S. 32, *33*
Turkey, Skoda 149mm vz 37 (K1) (**H**) 86

United States Army
  5th Army 67
  14th Quartermaster 189
American Expeditionary Force 41
  battalions
    2nd Ranger 114, *114*
    52nd Field Artillery 144
  divisions
    Third Infantry Division 196
  regiments
    11th Field Artillery 42
    119th Field Artillery 42
    129th Field Artillery 32
    165th Infantry 113
United States guns
  American Civil War
    3in ordnance rifle 12
    8in 200-pdr cannon 17, *18*
    10-pdr Parrott rifle 12
    12-pdr 1857 (**H**) 11, *12*, 12
    12-pdr Whitworth rifle 12–13
    13in mortar 'Dictator' 17, *17*
    18-pdr rifle 'Whistling Dick' 17
    20-pdr Parrott rifle 12
    24-pdr (**H**) 12
  World War I
    8in (**H**) 41
    9.2in (**H**) 41
    18-pounder 41
    35.6cm (14in) **RG** 60
    37mm Model 1916 41

    75mm 1897 (**FG**) 29, 32, *32*, 41
    120mm Model 1906 41
    155mm GPF 41, 42
    155mm Schneider (**H**) 41
    240mm (**H**) 41
  interwar period
    8in M1 (**H**) **68**
    155mm pack M1 (**H**) 66–7
    155mm M1918 (**H**) 66
  World War II
    0.50 cal. Maxson Mount 109, *111*
    3in M3 (**AA**) 109
    3in Model M5 (**ATG**) 120–1
    8in gun 67
    20mm Aut. Gun Mk IV (**AA**) 109
    37mm M1 (**AA**) 109–10
    37mm M1A2 (**AA**) 110
    37mm M3 (**ATG**) *67*, 120
    37mm Combination Mount M54 (**AA**) 110
    40mm Gun M1 (**AA**) 104, *105*
    57mm M1 (**ATG**) 117
    75mm pack M1A1 (**H**) 66, *70*
    90mm Gun M1 (**AA**) 109, *110*
    90mm Gun M1A1 (**AA**) 109
    105mm (**H**) M2A1 65–6, **94**
    105mm (**H**) M3 65
    105mm M7 Priest (**SP**) 126, **129**
    155mm M1 Long Tom 67, *70*
    155mm M40 (**SP**) 126–7
    203mm M43 (**SP**) 127
    240mm M1 (**H**) 67
    914mm mortar (Little David) 139
  Cold War
    75mm M20 (**RW**) 165
    105mm L118/L119 (**H**) 144, 146
    105mm M2A1 (**H**) 144, *145*
    105mm M7 Priest (**SP**) 153
    105mm M40 (**RW**) 165
    105mm M50 Ontos (**RW**) 165, *166*
    105mm M52 (**SP**) 154
    105mm M102 (**H**) 144, *145*
    105mm M108 (**SP**) 153–4
    155mm M2 Long Tom 144
    155mm M40 (**SP**) 153, *154*
    155mm M44 (**SP**) 154, *155*

    155mm M52 (**SP**) 154–5, *155*
    155mm M109 (**SP**) 154, **158**
    155mm M114 (**H**) 144
    155mm M198 (**GH**) 144–5
    175mm M107 (**SP**) 154
    203mm M110 (**SP**) 154, 187, *188*
    203mm M115 (**H**) 144
    240mm M1 (**H**) 144
    280mm 'Atomic Annie' 151, *152*
  modern artillery
    20mm M-163 (**SP**) *189*
    20mm Phalanx (**SP**) 210–11
    155mm M109A6 Paladin (**SP**) 196, *196*
    155mm M198 (**H**) *186*, *197*
United States missiles
  BGM-71 TOW/2 (**ATM**) 167–8
  CKEM (**ATM**) 214
  FGM-148 Javelin (**ATM**) 168,
    213–14, **216**
  FIM-92 Stinger (**AA**) 184
  M47 Dragon/II/Super Dragon (**ATM**) 168,
    169, **170**
  MGM-2 Lance 191
  MGM-5 Corporal 191
  MGM-29 Sergeant 191, 194
  MGM-52 Lance 194
  MGM-140 ATACMS 194
  MGR-1 Honest John 191
  MIM-23 HAWK (**AA**) 182
  MIM-72/M48 Chaparral (**AA**) 183, *183*
  MIM-104 Patriot (**AA**) 182, 183, *184*,
    188–90, *190*, 210
  Nike Ajax (**AA**) 182
  Nike Hercules (**AA**) 182
United States rockets
  4.5in M8 134
  4.5in M16 134
  HIMARS 212–13
  M270 MLRS 173, **177**, 187, *215*
USAF
  Eighth US Air Force 101
    squadrons
      354th Tactical Fighter 175
      578th Bomb 101
    US Marines

    1st Marine Raider Battalion 65
    Mike Battery 414 196
    regiments
      10th Marines 196
      11th Marines 65–6

Versailles, Treaty of 48, 53, 78–9, 103
Vietnam War 144, 150, 153–4, 155, 163, 165,
  167, 175, 178, 179, 196

Warsaw Pact 141, 150–1, 155
Warsaw Uprising 125, 126
Winter War 73, 121
World War I
  attack on Paris 59–61
  Battle of Le Cateau 34, 35–7
  Belgium invaded 43, 46, 54
  Dardanelles campaign *38*, 39
  First Battle of Ypres 35
  Marne battles 29, 42
  Somme offensive 40–1
  Verdun offensive 29, 33, 40, 54–6, 60
  Vimy Ridge offensive 39–40
World War II
  Anzio landings 134, 138
  Battle of the Bulge 109, 126
  Battle of Guadalcanal 65–6
  Battle of Kursk 115–16, *115*
  Battle of Stalingrad 74–5, *75*, 139
  Italian Campaign 117, 120, 126
  North Africa 67, 70, 79, 80–1, 90, 117
  Operation Barbarossa 79, 116, 130
  Pacific war 109, 111–13, 120
  Poland invaded 79, 96

Yom Kippur war 146, 161, 167, 182, 196
Ypres 46
Yugoslavia
  interwar period
    149mm Skoda vz 37 (K1) (**H**) 86
    220mm Skoda M28 (**H**) 87

Zhukov, Marshal Georgi 99